Conceptual Kinematics

A Companion to I. E. Irodov's Problems in General Physics

By

Chandra Shekhar Kumar
Integrated M. Sc. in Physics, IIT Kanpur, India
CEO & Co-Founder, Ancient Kriya Yoga Mission
CTO & Co-Founder, Ancient Science Publishers

Ancient Science Publishers

TeX is a trademark of the American Mathematical Society.

METAFONT is a trademark of Addison-Wesley.

Care has been taken in the preparation of this book, but makes no expressed or implied warranty of any kind and assumes no responsibility for errors or omissions. No liability is assumed for incidental or consequential damages in connection with or arising out of the use of the information contained herein.

For comments, suggestions and or feedback, send mail to :
ancientsciencepublishers@gmail.com

Copyright ©2018 Chandra Shekhar Kumar

All rights reserved. This work is protected by copyright and permission must be obtained prior to any prohibited reproduction, storage in a retrieval system, or transmission in any form or by any means, electronic, mechanical, photocopying, recording, or likewise unless stated otherwise.

ISBN-13: 978-1985065598
ISBN-10: 1985065592

Ancient Science Publishers, # 4, Suresh Prasad Singh, Sardar Patel Path, Boring Road, Patna, Bihar 800013, India.

Highly Recommended Books for Self Study & Competitions

Author : Chandra Shekhar Kumar

Conceptual Geometry of Straight Line
A Companion to S. L. Loney's Co-ordinate Geometry

Conceptual Geometry
A Companion to S. L. Loney's Co-ordinate Geometry

Conceptual Trigonometry Part I
A Companion to S. L. Loney's Plane Trigonometry Part I

Conceptual Trigonometry Part II
A Companion to S. L. Loney's Plane Trigonometry Part II

Conceptual Dynamics
A Companion to S. L. Loney's Elements of Dynamics

Conceptual Statics
A Companion to S. L. Loney's Elements of Statics

Conceptual Particle Dynamics
A Companion to S. L. Loney's Dynamics of A Particle

Conceptual Rigid Body Dynamics
A Companion to S. L. Loney's Dynamics of Rigid Bodies

Conceptual School Geometry
A Companion to Hall & Stevens' School Geometry

Conceptual School Algebra
A Companion to Hall & Knight's Elementary Algebra

Problems and Solutions in Plane Trigonometry
by
Isaac Todhunter & Neeru Singh

Solutions of the Examples in Higher Algebra
by
H. S. Hall, S. R. Knight, Neeru Singh
& C. S. Kumar

Questions and Problems in School Physics
A Companion to I. E. Irodov's Problems in General Physics
by
Lev Tarasov, Aldina Tarasova
& Chandra Shekhar Kumar

Calculus
Basic Concepts for High Schools
by
Lev Tarasov & Chandra Shekhar Kumar

General Methods for Solving Physics Problems
A Companion to I. E. Irodov's Problems in General Physics
by
B. S. Belikov & Chandra Shekhar Kumar

Preface

Back in 1990, Solving **Problems in General Physics** by **I. E. Irodov** had a terrorizing effect on me, irrespective of the outcome of the countless hours, full of perspiration and inspiration, laced with joy and surrendering to the sheer beauty and elegance of each problem, sub-problem, ... woven with multi-concepts.. Whenever stuck, I used to rely on the tomes of **Feynman Lectures on Physics**... an irresistible journey... back n forth between the classics of Irodov and Feynman.

As time grew, I ended up stocking a huge pile of sheets comprising of my notes as an endeavor to solve and devour the entire book and beyond (needless to mention that I laid my hands on everything I could in my pursuit).

Somewhere in 2005, I started collating and organizing my notes to instill coherence and capture the elegance in the flow.

The present work is an outcome of this pursuit, which will serve as a complete guide to private students reading the subject with few or no opportunities of instruction. This will save the time and lighten the work of Teachers as well. This book helps in acquiring a better understanding of the basic principles of Kinematics and in revising a large amount of the subject matter quickly. Care has been taken, as in the forthcoming ones, to present the solutions with multi-concepts and beyond in a simple natural manner, in order to meet the difficulties which are most likely to arise, and to render the work intelligible and instructive.

This work contains several variations of problems, solutions, methods, approaches related to analytics, graphical geometry, calculus, trigonometric geometry, scalar/vector algebra, differential equations, extrema without calculus to enrich, strengthen and enliven the inherent multi-concepts.

Ancient Science Publishers *Chandra Shekhar Kumar*
January, 2018.

List of Chapters

Preface   i

1 Classic Chase Problem and Curves of Pursuit   **1**
    1.1 Linear Pursuit 1
        1.1.1 Rectilinear Interception 1
            1.1.1.1 Angle Bisection 4
            1.1.1.2 Side Division 5
            1.1.1.3 Distance Computation 6
        1.1.2 Curvilinear Pure Pursuit 8
            1.1.2.1 Using Vector Equations 8
            1.1.2.2 Using Differential Equations 10
    1.2 Cyclic Pursuit 20
    1.3 Triangular Pursuit 23

2 Motorboat Problem and Crossing The River   **27**
    2.1 Rendezvous with Floating Raft 27
    2.2 Rendezvous with Anchored Ports 31
    2.3 Crossing The River 38

Chapter 1

Classic Chase Problem and Curves of Pursuit

1.1 Linear Pursuit

1.1.1 Rectilinear Interception

§ Problem 1.1.1. *A ghost, spell-bound by a tantrik (ghost-catcher), is confined to move on a fixed straig-ht line AB with a constant speed v_g. At the same time, the tantrik, situated at a point T, which is at a perpendicular distance D from AB, moves with a fixed speed of v_T in a direction to catch the ghost if it can or minimize his distance from the ghost.*

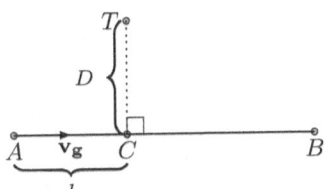

Assuming no knowledge of calculus and $AC = l$ as shown in the figure, help him determine his direction and duration of the chase including the minimum distance from the ghost.

◊

§§ Solution. Let N and L be the positions of the tantrik and the ghost respectively when they are nearest possible to each other. Let a circle be drawn with T as the center and TN as the radius.

Then the distance NL, being the least possible, the point N must lie on the straight line TL, joining the point L and the center T, because no other point in the whole periphery, at which the tantrik from T might arrive in the same time, is so near to L as that wherein the line TL intersects the said periphery.

1.1. Linear Pursuit

To compute the distance NL, a line TC perpendicular to the line AB is drawn such that $AC = l$, $CL = x$, $TC = D$.

$$\therefore TL = \sqrt{TC^2 + CL^2}$$
$$= \sqrt{D^2 + x^2} \qquad (1.1)$$

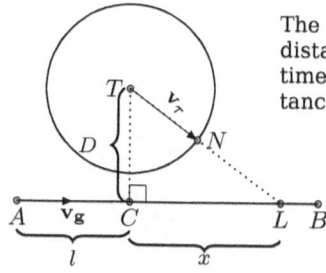

The time taken by the tantrik to cover the distance TN at speed v_τ is equal to the time taken by the ghost to cover the distance AL at speed v_g.

$$\therefore \frac{TN}{v_\tau} = \frac{AL}{v_g}$$
$$\therefore TN = (l+x)\frac{v_\tau}{v_g}. \qquad (1.2)$$

Eq. (1.1) - Eq. (1.2) $\implies$

$$NL = TL - TN = \sqrt{D^2 + x^2} - (l+x)\frac{v_\tau}{v_g} \qquad (1.3)$$

Let this minimum NL be denoted by q.

$$\therefore \text{Eq. (1.3)} \implies \sqrt{D^2 + x^2} - x\frac{v_\tau}{v_g} = q + l\frac{v_\tau}{v_g} \qquad (1.4)$$

Since q is minimum and $l\frac{v_\tau}{v_g}$ is a constant, hence

$$r = q + l\frac{v_\tau}{v_g} \qquad (1.5)$$

is a minimum.

Substituting r from Eq. (1.5) into Eq. (1.4):

$$\sqrt{D^2 + x^2} - x\frac{v_\tau}{v_g} = r \qquad (1.6)$$

$$\therefore D^2 + x^2 = \left(r + x\frac{v_\tau}{v_g}\right)^2 = r^2 + x^2\frac{v_\tau^2}{v_g^2} + 2rx\frac{v_\tau}{v_g}$$

$$\therefore \left(1 - \frac{v_\tau^2}{v_g^2}\right)x^2 - 2rx\frac{v_\tau}{v_g} = r^2 - D^2$$

$$\therefore x^2 + 2\frac{rv_\tau v_g}{v_\tau^2 - v_g^2}x + \frac{v_g^2\left(r^2 - D^2\right)}{v_\tau^2 - v_g^2} = 0 \qquad (1.7)$$

Solving this quadratic equation:

$$x = \frac{rv_\tau v_g}{v_g^2 - v_\tau^2} \pm \sqrt{\left(\frac{rv_\tau v_g}{v_g^2 - v_\tau^2}\right)^2 + \frac{v_g^2\left(r^2 - D^2\right)}{v_g^2 - v_\tau^2}}$$

$$= \frac{rv_\tau v_g \pm \sqrt{r^2 v_\tau^2 v_g^2 + \left(v_g^4 - v_g^2 v_\tau^2\right)\left(r^2 - D^2\right)}}{v_g^2 - v_\tau^2}$$

$$\therefore x = \frac{rv_\tau v_g \pm \sqrt{v_g^4 r^2 - D^2 v_g^2\left(v_g^2 - v_\tau^2\right)}}{v_g^2 - v_\tau^2} \qquad (1.8)$$

Kindly note that when $v_g < v_\tau$, the term $-D^2 v_g^2\left(v_g^2 - v_\tau^2\right)$ becomes a positive quantity, therefore there remains no condition of r becom-

1.1. Linear Pursuit

ing a minimum.
$$\therefore v_g > v_\tau \tag{1.9}$$

Now it is evident that $v_g^4 r^2$ or r cannot be taken so small as to make the root impossible.

Therefore, when r is minimum, we must have
$$v_g^4 r^2 = D^2 v_g^2 \left(v_g^2 - v_\tau^2\right)$$
$$\therefore r = \frac{D\sqrt{v_g^2 - v_\tau^2}}{v_g} \tag{1.10}$$

For this minimum of r, the Eq. (1.8) yields the following value of x:
$$\therefore x = \frac{r v_g v_\tau}{v_g^2 - v_\tau^2} = D \frac{v_\tau}{\sqrt{v_g^2 - v_\tau^2}} \tag{1.11}$$

From Eq. (1.3):
$$NL = \sqrt{D^2 + x^2} - (l+x)\frac{v_\tau}{v_g}$$
$$= \sqrt{D^2 + x^2} - x\frac{v_\tau}{v_g} - l\frac{v_\tau}{v_g}$$
$$= r - l\frac{v_\tau}{v_g} \text{ (From Eq. (1.6))}$$
$$= \frac{D\sqrt{v_g^2 - v_\tau^2}}{v_g} - l\frac{v_\tau}{v_g} \text{ (From Eq. (1.10))}$$
$$\therefore NL = \frac{D\sqrt{v_g^2 - v_\tau^2} - l v_\tau}{v_g}. \tag{1.12}$$

It is evident that for NL to be a positive measurable quantity, From Eq. (1.9) and Eq. (1.12):
$$v_g > v_\tau \ \& \ D\sqrt{v_g^2 - v_\tau^2} > l v_\tau.$$

Duration of the chase is given by (From Eq. (1.2)):
$$\tau = \frac{TN}{v_\tau} = \frac{AL}{v_g}$$
$$\therefore \tau = \frac{l+x}{v_g}$$
$$\therefore \tau = \left(\frac{l}{v_g} + D\frac{v_\tau}{v_g\sqrt{v_g^2 - v_\tau^2}}\right). \text{ (From Eq. (1.11))}$$

Alternative Solution to Eq. (1.7):

$$\text{Let } A = \frac{r v_\tau v_g}{v_g^2 - v_\tau^2} \tag{1.13}$$
$$B = \frac{v_g^2 \left(r^2 - D^2\right)}{v_g^2 - v_\tau^2} \tag{1.14}$$

Then Eq. (1.7) becomes
$$x^2 - 2Ax = B$$

Let $x = A + y$
$$\therefore A^2 + y^2 + 2Ay - 2A^2 - 2Ay = B$$
$$\therefore y^2 - A^2 = B$$
$$\therefore y^2 = A^2 + B \tag{1.15}$$

1.1. Linear Pursuit

Substituting the values of A and B from Eq. (1.13) and Eq. (1.14) respectively into Eq. (1.15):

$$\therefore y^2 = \left(\frac{r v_\tau v_g}{v_g^2 - v_\tau^2}\right)^2 + \frac{v_g^2 \left(r^2 - D^2\right)}{v_g^2 - v_\tau^2}$$

$$\therefore y^2 \left(v_g^2 - v_\tau^2\right)^2 = r^2 v_g^2 v_\tau^2 + \left(v_g^4 - v_g^2 v_\tau^2\right)\left(r^2 - D^2\right)$$

$$\therefore r^2 = \frac{y^2 \left(v_g^2 - v_\tau^2\right)^2 + D^2 v_g^2 \left(v_g^2 - v_\tau^2\right)}{v_g^4} \tag{1.16}$$

The above is minimum when $y^2 = 0$, i.e.

$$\therefore r = D \frac{\sqrt{v_g^2 - v_\tau^2}}{v_g} \tag{1.17}$$

which is the same as Eq. (1.10).

$$\therefore x = A + y = A = \frac{r v_\tau v_g}{v_g^2 - v_\tau^2} = D \frac{v_\tau}{\sqrt{v_g^2 - v_\tau^2}}$$

which is the same as Eq. (1.11). ∎

§ **Problem 1.1.2.** *A ghost, situated at a point G, spell-bound by a tantrik (ghost-catcher), is confined to move on a fixed straight line with a constant speed v_g. At the same time, the tantrik, situated at a point T, moves with a fixed speed of v_τ in a direction to catch the ghost in the least possible time. Assuming $GT = l$, determine the tantrik's course in pursuit.* ◊

§§ **Solution.** Assume $\mu = \dfrac{v_\tau}{v_g} > 1$, it is possible for the tantrik to take a straight course such that it catches the ghost in the least possible time.

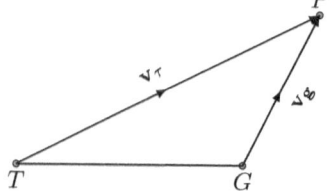

P is the point at which the tantrik T moving in a straight line TP meets the ghost G moving in a straight line GP. Note that

$$\frac{TP}{GP} = \mu = \frac{v_\tau}{v_g}. \tag{1.18}$$

Finding all the points which can be simultaneously reached by both the tantrik and the ghost for a given α will yield a circle as can be deduced ahead.

1.1.1.1 Angle Bisection

PL is the internal bisector of the $\angle TPG$.

$$\therefore \beta = \alpha \tag{1.19}$$

PR is the external bisector of the angle at P.

$$\therefore \gamma = \delta \tag{1.20}$$

Adding Eq. (1.19) and Eq. (1.20):

$$\beta + \gamma = \alpha + \delta \tag{1.21}$$

$$\text{But} \quad \alpha + \beta + \gamma + \delta = \pi \tag{1.22}$$

1.1. Linear Pursuit

Hence Eq. (1.21) and Eq. (1.22) $\implies \beta + \gamma = \frac{\pi}{2} = \angle LPR$.

GQ is drawn parallel to LP.
$\therefore \angle GMR = \angle LPR = \angle LPM$.
In $\triangle GPM$ and $\triangle QPM$: the side PM is common and
$$\angle GPM = \gamma = \delta = \angle QPM,$$
$$\angle PMG = \angle PMQ = \frac{\pi}{2}.$$
$$\therefore \triangle GPM \cong \triangle QPM$$

$$\therefore PQ = PG. \tag{1.23}$$

In $\triangle TGQ$:
$$\because LP \parallel GQ,$$
$$\therefore \frac{TL}{LG} = \frac{TP}{PQ} = \frac{TP}{PG} \text{ (Using Eq. (1.23))}.$$
$$\therefore \frac{TL}{LG} = \mu \text{ (Using Eq. (1.18))} \tag{1.24}$$

Eq. (1.24) $\implies$ The bisector of the interior angle of a triangle divides the opposite side into parts proportional to the adjacent sides.

And, the same holds for the exterior angle as well, i.e.
$$\therefore \frac{TR}{GR} = \frac{TP}{PG} = \mu. \tag{1.25}$$

Eq. (1.24) and Eq. (1.25) $\implies$
$$\frac{TL}{LG} = \frac{TR}{GR} = \mu. \tag{1.26}$$

$\because$ Eq. (1.26) doesn't depend on P, hence for a given μ, the same points L and R will be found for a new point P. For every such point P, $\angle LPR = \frac{\pi}{2}$, hence the locus of P is a semi-circle with LR as the diameter. With P on the other side of TG, other side of the circle can be found. It can be noticed that the point G is not the center of this circle.

If this circle cuts the course of the ghost, the tantrik has to move directly towards the point of intersection to catch the ghost in the least possible time; if it doesn't cut the course then it is not possible to catch the ghost.

1.1.1.2 Side Division

Let L and R be the points which divides the side TG of the $\triangle TPG$ internally and externally in the same ratio as that of the sides TP and GP, i.e.
$$\frac{TL}{LG} = \frac{TR}{GR} = \frac{PT}{PG} = \mu.$$
$$\therefore \left|\vec{PT}\right| = \mu \left|\vec{PG}\right|$$
$$\therefore \left|\vec{PT}\right|^2 = \mu^2 \left|\vec{PG}\right|^2$$

$$\therefore \left(\mu|\vec{PG}| + |\vec{PT}|\right) \cdot \left(\mu|\vec{PG}| - |\vec{PT}|\right) = 0$$

$$\therefore \left(\frac{\mu|\vec{PG}| + |\vec{PT}|}{\mu + 1}\right) \cdot \left(\frac{\mu|\vec{PG}| - |\vec{PT}|}{\mu - 1}\right) = 0$$

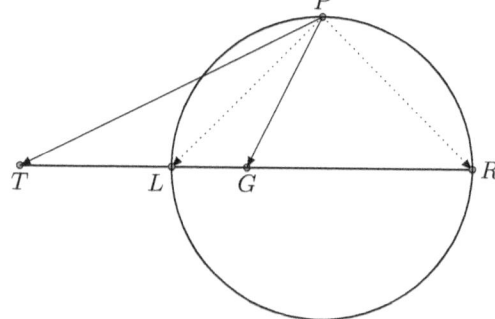

$$\therefore \vec{PL} \cdot \vec{PR} = 0$$
$$\therefore \vec{PL} \perp \vec{PR}$$
$$\therefore \angle LPR = \frac{\pi}{2}.$$

Hence, the point P is on a circle with its diameter as LR.

1.1.1.3 Distance Computation

Setting up a rectangular co-ordinate system with $T(0,0)$ as the origin, TR as the y-axis, $P(x,y)$ and $G(l,0)$, the distance computation can be done as :

$$PT = \mu PG$$
$$\therefore PT^2 = \mu^2 PG^2$$
$$\therefore x^2 + y^2 = \mu^2 \left[(x-l)^2 + y^2)\right]$$
$$\therefore \left(\mu^2 - 1\right)x^2 - 2l\mu^2 x + \left(\mu^2 - 1\right)y^2 + l\mu^2 = 0$$
$$\therefore \left(x - \frac{\mu^2 l}{\mu^2 - 1}\right)^2 + y^2 = \left(\frac{\mu l}{\mu^2 - 1}\right)^2 \quad (1.27)$$

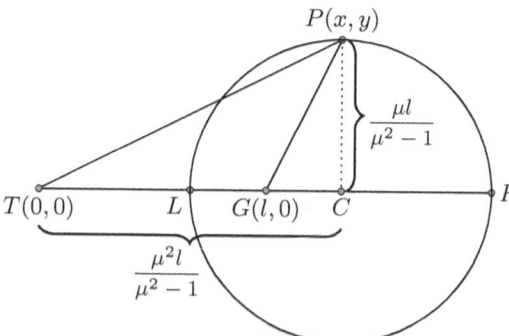

Eq. (1.27) is an equation of a circle with its center at $C\left(\frac{\mu^2 l}{\mu^2 - 1}, 0\right)$ and radius of $\frac{\mu l}{\mu^2 - 1}$. Note that $\mu \neq 1$.

If $\mu < 1$, it is still possible for the slow tantrik to catch the fast ghost.

Suppose that the ghost heads in such a direction θ to GT that it is tangent to the circle at a point Q. $\therefore \angle CQG = \frac{\pi}{2}$.

Note that if the absolute value of the heading angle of the ghost $> \theta$, then no interception is possible, and if it is $< \theta$, then the tantrik's

1.1. Linear Pursuit

path will cross the circle twice and so there will be two possible interception points Q.

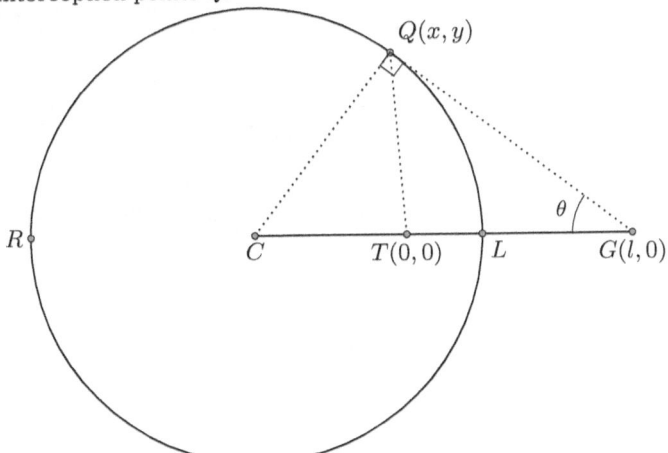

Since, $\dfrac{TQ}{GQ} = \mu$, hence the locus of Q is a circle and the equation is given by (as in Eq. (1.27), note that $\mu < 1$):

$$\therefore \left[x - \left(-\frac{\mu^2 l}{1-\mu^2}\right)\right]^2 + y^2 = \left(\frac{\mu l}{1-\mu^2}\right)^2 \qquad (1.28)$$

In $\triangle CQG$:

$$\sin\theta = \frac{CQ}{CG} = \frac{CQ}{CT + TG}$$

$$= \frac{\dfrac{\mu l}{1-\mu^2}}{\dfrac{\mu^2 l}{1-\mu^2} + l} = \mu = \frac{v_\tau}{v_g}$$

$$\therefore \theta = \sin^{-1}\frac{v_\tau}{v_g}. \qquad (1.29)$$

Hence for $\mu < 1$, the interception is possible only when $\theta \leq \sin^{-1}\dfrac{v_\tau}{v_g}$.

Note that if β is the heading angle of the tantrik, then the interception is possible only when $-\beta \leq \theta \leq \beta$, impossible otherwise.

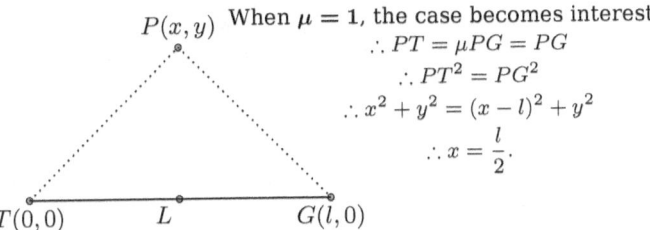

When $\mu = 1$, the case becomes interesting.

$$\therefore PT = \mu PG = PG$$
$$\therefore PT^2 = PG^2$$
$$\therefore x^2 + y^2 = (x-l)^2 + y^2$$
$$\therefore x = \frac{l}{2}.$$

So the locus of P is a straight line which is a perpendicular bisector of TG, which is a circle of infinite radius in principle. ∎

1.1. Linear Pursuit

1.1.2 Curvilinear Pure Pursuit

§ Problem 1.1.3. *A ghost, situated at a point G, spell-bound by a tantrik (ghost-catcher), is confined to move on a fixed straight line GR with a constant speed v_g.*

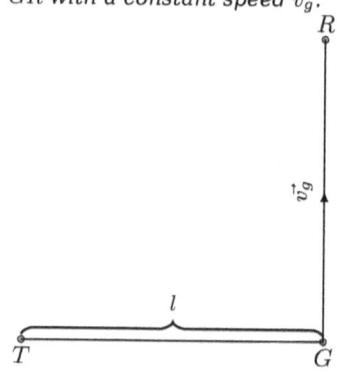

At the same time, the tantrik, situated at a point T, which is at a perpendicular distance l from GR as shown in the figure, always moves directly towards the ghost with a fixed speed of v_τ to catch the ghost if it can or minimize his distance from the ghost.

Determine the duration of the chase τ_d, point of intersection (if caught), length of the curve of pursuit and equation of the tantrik's trajectory. ◊

§§ Solution. Setting up a rectangular co-ordinate system with $T(0,0)$ as the origin, TG as the x-axis, TY as the y-axis and $G(l,0)$, the computation becomes a bit easier.

It is given that the instantaneous velocity of the tantrik is always directed towards the instantaneous position of the ghost.

1.1.2.1 Using Vector Equations

Let $\vec{r_\tau}$ and $\vec{r_g}$ be the radius vectors of the tantrik and ghost respectively.

Let T_t and G_t be the instantaneous positions of the tantrik and ghost respectively at time t.

It is given that

$$\frac{d\vec{r_\tau}}{dt} = \vec{v_\tau} \tag{1.30}$$

$$\frac{d\vec{r_g}}{dt} = \vec{v_g} \tag{1.31}$$

Because the tantrik is directed always towards the ghost, hence

$$\vec{r_g} - \vec{r_\tau} \propto \vec{v_\tau}$$
$$\therefore \vec{r_g} - \vec{r_\tau} = \gamma(t)\vec{v_\tau} \tag{1.32}$$

where $\gamma(t)$ is the proportionality coefficient which depends on time.

At the beginning of the chase, i.e. at $t = 0$: $|\vec{r_g} - \vec{r_\tau}| = TG = l$;

$$\therefore \gamma(0) = \frac{l}{v_\tau} \tag{1.33}$$

At the end of the chase, i.e. at $t = \tau_d$: $|\vec{r_g} - \vec{r_\tau}| = 0$;

$$\therefore \gamma(\tau_d) = 0 \tag{1.34}$$

Because the tantrik moves with a constant speed of v_τ;

$$\therefore \frac{dv_\tau^2}{dt} = 0$$

$$\therefore \frac{d(\vec{v_\tau} \cdot \vec{v_\tau})}{dt} = 0 = \frac{d\vec{v_\tau}^2}{dt}$$

$$\therefore 2\vec{v_\tau} \cdot \frac{d\vec{v_\tau}}{dt} = 0$$

1.1. Linear Pursuit

$$\therefore \vec{v_\tau} \cdot \frac{d\vec{v_\tau}}{dt} = 0 \qquad (1.35)$$

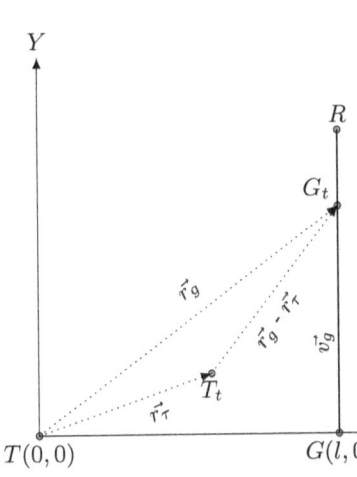

From Eq. (1.35), it is evident that the tantrik's velocity $\vec{v_\tau}$ is always perpendicular to his acceleration $\vec{a_\tau} = \frac{d\vec{v_\tau}}{dt}$. $\therefore \vec{v_\tau} \perp \vec{a_\tau}$.

Since the ghost is confined to move in a fixed (direction) straight line GR with a constant speed v_g,

$$\therefore \frac{d\vec{v_g}}{dt} = 0 \qquad (1.36)$$

Differentiating both sides of Eq. (1.32) w.r.t. time,

$$\frac{d}{dt}(\vec{r_g} - \vec{r_\tau}) = \frac{d}{dt}(\gamma(t)\vec{v_\tau})$$

$$\therefore \frac{d\vec{r_g}}{dt} - \frac{d\vec{r_\tau}}{dt} = \gamma(t)\frac{d\vec{v_\tau}}{dt} + \vec{v_\tau}\frac{d\gamma(t)}{dt}$$

$$\therefore \vec{v_g} - \vec{v_\tau} = \gamma(t)\frac{d\vec{v_\tau}}{dt} + \vec{v_\tau}\frac{d\gamma(t)}{dt} \qquad (1.37)$$

Taking dot product of both sides with $\vec{v_\tau}$,

$$\therefore \vec{v_\tau} \cdot (\vec{v_g} - \vec{v_\tau}) = \vec{v_\tau} \cdot \left(\gamma(t)\frac{d\vec{v_\tau}}{dt} + \vec{v_\tau}\frac{d\gamma(t)}{dt}\right)$$

$$\therefore \vec{v_\tau} \cdot \vec{v_g} - \vec{v_\tau} \cdot \vec{v_\tau} = \gamma(t)\vec{v_\tau} \cdot \frac{d\vec{v_\tau}}{dt} + \vec{v_\tau} \cdot \vec{v_\tau}\frac{d\gamma(t)}{dt}$$

$$\therefore \vec{v_g} \cdot \vec{v_\tau} - v_\tau^2 = v_\tau^2 \frac{d\gamma(t)}{dt} \text{ [Using Eq. (1.35)]}$$

$\because \vec{v_g} = v_g\,\hat{\mathbf{j}}$, Assuming $\vec{v_\tau} = v_\tau^x\,\hat{\mathbf{i}} + v_\tau^y\,\hat{\mathbf{j}}$,

$$\therefore v_g\,v_\tau^y - v_\tau^2 = v_\tau^2 \frac{d\gamma(t)}{dt} \qquad (1.38)$$

Integrating both sides of Eq. (1.38) w.r.t. time,

$$v_g \int_0^{\tau_d} v_\tau^y\,dt - v_\tau^2 \int_0^{\tau_d} dt = v_\tau^2 \int_0^{\tau_d} \frac{d\gamma(t)}{dt}\,dt$$

$\because \int_0^{\tau_d} v_\tau^y\,dt$ is the distance traveled by the tantrik in y-direction in time τ_d which is the same as that traveled by the ghost in time $\tau_d = v_g \tau_d$

$$\therefore v_g\,(v_g\,\tau_d) - v_\tau^2\,\tau_d = v_\tau^2\,(\gamma(\tau_d) - \gamma(0))$$

(Using Eq. (1.33) & Eq. (1.34))

$$\therefore \tau_d\,(v_g^2 - v_\tau^2) = v_\tau^2 \left(0 - \frac{l}{v_\tau}\right)$$

$$\therefore \tau_d = \frac{l v_\tau}{v_\tau^2 - v_g^2} \qquad (1.39)$$

Alternatively: Instead of taking the route of Eq. (1.37) and Eq. (1.38), the derivation of Eq. (1.39) can be done as follows.

$$\frac{d}{dt}\left[(\vec{r_g} - \vec{r_\tau}) \cdot (\vec{v_g} + \vec{v_\tau})\right]$$

1.1. Linear Pursuit

$$= (\vec{r_g} - \vec{r_\tau}) \cdot \frac{d}{dt}(\vec{v_g} + \vec{v_\tau}) + (\vec{v_g} + \vec{v_\tau}) \cdot \frac{d}{dt}(\vec{r_g} - \vec{r_\tau})$$

$$= (\vec{r_g} - \vec{r_\tau}) \cdot \left(\frac{d\vec{v_g}}{dt} + \frac{d\vec{v_\tau}}{dt}\right) + (\vec{v_g} + \vec{v_\tau}) \cdot (\vec{v_g} - \vec{v_\tau})$$

$$= (\vec{r_g} - \vec{r_\tau}) \cdot \frac{d\vec{v_\tau}}{dt} + \left(v_g^2 - v_\tau^2\right) \text{ [Using Eq. (1.36)]}$$

$$= \gamma(t)\vec{v_\tau} \cdot \frac{d\vec{v_\tau}}{dt} + \left(v_g^2 - v_\tau^2\right) \text{ [Using Eq. (1.32)]}$$

$$= v_g^2 - v_\tau^2 = \text{ a constant [Using Eq. (1.35)]}$$

Integrating both sides w.r.t. time from $t = 0$ to $t = \tau_d$,

$$[(\vec{r_g} - \vec{r_\tau}) \cdot (\vec{v_g} + \vec{v_\tau})]\bigg|_{(t=\tau_d)} - [(\vec{r_g} - \vec{r_\tau}) \cdot (\vec{v_g} + \vec{v_\tau})]\bigg|_{(t=0)}$$

$$= \int_0^{\tau_d} \left(v_g^2 - v_\tau^2\right) dt$$

$$\therefore 0 - l\hat{\mathbf{i}} \cdot \left(v_g\hat{\mathbf{j}} + v_\tau\hat{\mathbf{i}}\right) = \left(v_g^2 - v_\tau^2\right)\tau_d$$

$$\therefore \tau_d = \frac{lv_\tau}{v_\tau^2 - v_g^2}$$

Distance traveled by the ghost in time τ_d is given by

$$v_g\tau_d = \frac{lv_\tau v_g}{v_\tau^2 - v_g^2}$$

Hence, the point of intersection is

$$\left(l, \frac{lv_\tau v_g}{v_\tau^2 - v_g^2}\right)$$

Length of the curve of pursuit by the tantrik is the same as the distance traveled by him in time τ_d which is given by

$$v_\tau \tau_d = \frac{lv_\tau^2}{v_\tau^2 - v_g^2}$$

1.1.2.2 Using Differential Equations

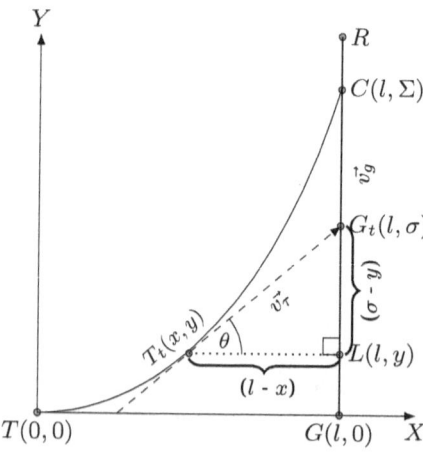

Because the tantrik is directed always towards the ghost, hence the trajectory of his path may look like the curve TT_tC, of which ds is an infinitesimal element, where $C(l, \Sigma)$ is the point of capture or intersection. $T_t(x, y)$ is the position of the tantrik at time t, such that T_tG_t is a tangent to his path and $G_t(l, \sigma)$ is the ghost's position at the same time t.
Let $T_tG_t = \gamma$.
Arc length $TT_t = s$, Arc length $TT_tC = S$.

1.1. Linear Pursuit

$$\therefore \Sigma = GC = v_g \tau_d \tag{1.40}$$
$$S = v_\tau \tau_d \tag{1.41}$$
$$\therefore \frac{\Sigma}{S} = \frac{v_g}{v_\tau} = \beta = \text{ (a constant)} \tag{1.42}$$

Similarly
$$\therefore \frac{\sigma}{s} = \frac{v_g}{v_\tau} = \beta. \tag{1.43}$$
$$\therefore d\sigma = \beta \, ds \tag{1.44}$$

In right-angled $\Delta T_t L G_t$,
$$\sigma - y = \gamma \sin \theta \tag{1.45}$$
$$l - x = \gamma \cos \theta \tag{1.46}$$

Using Pythagoras Theorem,
$$T_t G_t^2 = T_t L^2 + L G_t^2$$
$$\therefore \gamma^2 = (l-x)^2 + (\sigma - y)^2$$
$$\therefore 2\gamma d\gamma = 2(l-x)(-dx) + 2(\sigma - y)(d\sigma - dy)$$
$$\therefore \gamma d\gamma = \gamma \sin \theta (d\sigma - dy) - \gamma \cos \theta dx$$
$$\therefore d\gamma = d\sigma \, \sin \theta - (dx \, \cos \theta + dy \, \sin \theta) \tag{1.47}$$

It is easy to see that
$$ds = dx \, \cos \theta + dy \, \sin \theta \tag{1.48}$$
$$dx = ds \cos \theta \tag{1.49}$$
$$dy = ds \sin \theta \tag{1.50}$$

Using Eq. (1.44), Eq. (1.48), Eq. (1.49) and Eq. (1.50), the Eq. (1.47) becomes
$$d\gamma = \beta \, ds \, \sin \theta - ds = \beta dy - ds$$
$$\therefore ds = \beta dy - d\gamma$$
$$\therefore \int_0^S ds = \beta \int_0^\Sigma dy - \int_l^0 d\gamma$$
$$\therefore S = \beta \Sigma + l = \beta^2 S + l \text{ [Using Eq. (1.42)]}$$
$$\therefore S = \frac{l}{1 - \beta^2}$$
$$\therefore S = \frac{l v_\tau^2}{v_\tau^2 - v_g^2} \text{ [Using Eq. (1.43)]}$$
$$\therefore \tau_d = \frac{S}{v_\tau} \text{ [Using Eq. (1.41)]}$$
$$\therefore \tau_d = \frac{l v_\tau}{v_\tau^2 - v_g^2}$$
$$\therefore \Sigma = v_g \tau_d \text{ [Using Eq. (1.40)]}$$
$$\therefore \Sigma = \frac{l v_g v_\tau}{v_\tau^2 - v_g^2}$$

$\therefore$ The point of intersection is $C(l, \Sigma) \equiv$
$$\left(l, \frac{l v_g v_\tau}{v_\tau^2 - v_g^2} \right)$$

Computation of path of trajectory, i.e. pursuit curve of the tantrik relative to the ghost is simple and simplified further in polar coordinates, taking the course of the ghost as the initial line $\theta = 0$.

1.1. Linear Pursuit

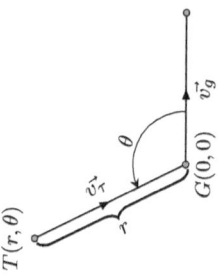

At time t, position of the tantrik is $T(r, \theta)$, position of the ghost is $G(0,0)$.

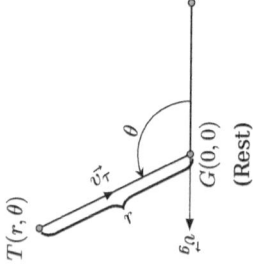

At time t, Motion of the tantrik $T(r, \theta)$ relative to the ghost $G(0,0)$ is as shown in the figure.

Resolving the tantrik's velocity in radial and transverse (tangential) directions:

$$\frac{dr}{dt} = -v_\tau + v_g \cos(\pi - \theta) = -(v_\tau + v_g \cos\theta) \tag{1.51}$$

$$r\frac{d\theta}{dt} = v_g \sin(\pi - \theta) = v_g \sin\theta \tag{1.52}$$

$Eq.\ (1.51) \div Eq.\ (1.52) \implies$

$$\frac{dr}{r} = -\left(\frac{v_\tau}{v_g}\csc\theta + \cot\theta\right)d\theta$$

$$\therefore \int_l^r \frac{dr}{r} = -\frac{v_\tau}{v_g}\int_{\frac{\pi}{2}}^\theta \csc\theta\, d\theta - \int_{\frac{\pi}{2}}^\theta \cot\theta\, d\theta$$

$$\therefore \ln\frac{r}{l} = -\frac{v_\tau}{v_g}\int_{\frac{\pi}{2}}^\theta \csc\theta\, d\theta - \int_{\frac{\pi}{2}}^\theta \cot\theta\, d\theta \tag{1.53}$$

$$\int \csc\theta\, d\theta = \int \frac{\csc\theta(\csc\theta + \cot\theta)}{\csc\theta + \cot\theta}\, d\theta \tag{1.54}$$

If $u = \csc\theta + \cot\theta$

Then $du = -\csc\theta \cot\theta\, d\theta - \csc^2\theta\, d\theta$

$$\therefore du = -\csc\theta(\csc\theta + \cot\theta)\, d\theta \tag{1.55}$$

Putting Eq. (1.55) into Eq. (1.54):

$$\therefore \int \csc\theta\, d\theta = -\int \frac{du}{u} = -\ln u = -\ln(\csc\theta + \cot\theta)$$

$$= -\ln\frac{1 + \cos\theta}{\sin\theta} = -\ln\frac{2\cos^2\frac{\theta}{2}}{2\sin\frac{\theta}{2}\cos\frac{\theta}{2}}$$

$$\therefore \int \csc\theta\, d\theta = -\ln\cot\frac{\theta}{2} = \ln\tan\frac{\theta}{2}. \tag{1.56}$$

1.1. Linear Pursuit

$$\int \cot\theta \, d\theta = \int \frac{\cos\theta}{\sin\theta} \, d\theta$$

If $u = \sin\theta$
Then $du = \cos\theta \, d\theta$

$$\therefore \int \cot\theta \, d\theta = \int \frac{du}{u} = \ln u = \ln\sin\theta \tag{1.57}$$

Putting Eq. (1.56) and Eq. (1.57) into Eq. (1.53):

$$\ln\frac{r}{l} = -\frac{v_\tau}{v_g}\ln\tan\frac{\theta}{2} - \ln\sin\theta$$

$$\therefore \ln\frac{l}{r} = \ln\left[\left(\tan\frac{\theta}{2}\right)^\mu \sin\theta\right], \text{ where } \mu = \frac{v_\tau}{v_g}.$$

$$\frac{l}{r} = \left(\tan\frac{\theta}{2}\right)^\mu \sin\theta$$

$$\therefore r = \frac{l}{\sin\theta}\left(\cot\frac{\theta}{2}\right)^\mu, \text{ where } \mu = \frac{v_\tau}{v_g}. \tag{1.58}$$

Thus from Eq. (1.58), it is easy to see that $r = 0$ when $\theta = \pi$.

Hence the time of pursuit τ_d, i.e., the time taken by the tantrik to capture the ghost can be computed as follows.

Using the Eq. (1.52) and Eq. (1.58):
($\because \sin\theta = 2\sin\frac{\theta}{2}\cos\frac{\theta}{2}$)

$$\int_0^{\tau_d} dt = \frac{l}{4v_g}\int_{\frac{\pi}{2}}^{\pi} \frac{\cos^{\mu-2}\frac{\theta}{2}}{\sin^{\mu+2}\frac{\theta}{2}} \, d\theta$$

$$\therefore \tau_d = \frac{l}{4v_g}\int_{\frac{\pi}{2}}^{\pi} \cot^{\mu-2}\frac{\theta}{2} \operatorname{cosec}^4\frac{\theta}{2} \, d\theta$$

$$= \frac{l}{4v_g}\int_{\frac{\pi}{2}}^{\pi} \cot^{\mu-2}\frac{\theta}{2}\left(1 + \cot^2\frac{\theta}{2}\right)\operatorname{cosec}^2\frac{\theta}{2} \, d\theta$$

Putting $\gamma = \cot\frac{\theta}{2}$

$$\therefore d\gamma = -\frac{1}{2}\operatorname{cosec}^2\frac{\theta}{2} \, d\theta$$

$$\therefore \tau_d = -\frac{l}{2v_g}\int_1^0 \left(\gamma^{\mu-2} + \gamma^\mu\right) d\gamma$$

$$= -\frac{l}{2v_g}\left(\frac{\gamma^{\mu-1}}{\mu-1} + \frac{\gamma^{\mu+1}}{\mu+1}\right)\Big|_1^0$$

$$= \frac{\mu l}{v_g(\mu^2 - 1)}, \text{ where } \mu = \frac{v_\tau}{v_g}$$

$$\therefore \tau_d = \frac{lv_\tau}{v_\tau^2 - v_g^2}. \tag{1.59}$$

As expected, this is the same as in Eq. (1.39).

For computation of path of trajectory, i.e. pursuit curve of the tantrik in the inertial frame of reference, Eq. (1.52) and Eq. (1.58)
$\implies$

$$\frac{l}{\sin\theta}\left(\cot\frac{\theta}{2}\right)^\mu \frac{d\theta}{dt} = v_g \sin\theta$$

$$\therefore \frac{4v_g}{l} \int_0^t dt = \int_{\frac{\pi}{2}}^{\theta} \frac{\cos^{\mu-2}\frac{\theta}{2}}{\sin^{\mu+2}\frac{\theta}{2}} d\theta = \int_{\frac{\pi}{2}}^{\theta} \cot^{\mu-2}\frac{\theta}{2} \operatorname{cosec}^4\frac{\theta}{2} d\theta$$

$$\therefore t = \frac{l}{4v_g} \int_{\frac{\pi}{2}}^{\theta} \cot^{\mu-2}\frac{\theta}{2} \left(1 + \cot^2\frac{\theta}{2}\right) \operatorname{cosec}^2\frac{\theta}{2} d\theta$$

Putting $\gamma = \cot\frac{\theta}{2}$

$$\therefore d\gamma = -\frac{1}{2} \operatorname{cosec}^2\frac{\theta}{2} d\theta$$

$$\therefore t = -\frac{l}{2v_g} \int_{\cot\frac{\pi}{4}}^{\cot\frac{\theta}{2}} \left(\gamma^{\mu-2} + \gamma^{\mu}\right) d\gamma$$

$$= -\frac{l}{2v_g} \left(\frac{\gamma^{\mu-1}}{\mu-1} + \frac{\gamma^{\mu+1}}{\mu+1}\right)\bigg|_1^{\cot\frac{\theta}{2}}$$

$$\therefore \frac{v_g(\mu^2-1)}{\mu l} t = 1 - \frac{1}{2\mu} \left\{(\mu+1)\cot^{\mu-1}\frac{\theta}{2} + (\mu-1)\cot^{\mu+1}\frac{\theta}{2}\right\} \quad (1.60)$$

Eq. (1.39) $\implies$

$$\tau_d = \frac{lv_\tau}{v_\tau^2 - v_g^2} = \frac{l\mu}{v_g(\mu^2-1)} \quad (1.61)$$

Eq. (1.61) and Eq. (1.60) $\implies$

$$\frac{t}{\tau_d} = 1 - \frac{1}{2\mu} \left\{(\mu+1)\cot^{\mu-1}\frac{\theta}{2} + (\mu-1)\cot^{\mu+1}\frac{\theta}{2}\right\} \quad (1.62)$$

Eq. (1.58) $\implies$

$$l - x = r\sin\theta = l\left(\cot\frac{\theta}{2}\right)^\mu$$

$$\therefore \cot\frac{\theta}{2} = \left(\frac{l-x}{l}\right)^{\frac{1}{\mu}} = \left(\frac{l-x}{l}\right)^\beta, \text{ where } \beta = \frac{1}{\mu} = \frac{v_g}{v_\tau} \quad (1.63)$$

Eq. (1.63) and Eq. (1.62) $\implies$

$$\frac{t}{\tau_d} = 1 - \frac{1}{2} \left\{(1+\beta)\left(\frac{l-x}{l}\right)^{1-\beta} + (1-\beta)\left(\frac{l-x}{l}\right)^{1+\beta}\right\} \quad (1.64)$$

From Eq. (1.38) and Eq. (1.33),

$$v_g v_\tau^y - v_\tau^2 = v_\tau^2 \frac{d\gamma(t)}{dt}$$

$$\therefore v_g \int_0^t v_\tau^y dt - v_\tau^2 \int_0^t dt = v_\tau^2 \left[\gamma(t) - \gamma(0)\right]$$

$$\therefore v_g y - v_\tau^2 t = \gamma(t) v_\tau^2 - l v_\tau$$

$$\therefore \frac{y}{l} = [t + \gamma(t)] \frac{v_\tau^2}{l v_g} - \frac{v_\tau}{v_g}$$

$$= [t + \gamma(t)] \frac{\mu v_\tau^2}{v_g^2 \tau_d (\mu^2 - 1)} - \mu, \text{ (From Eq. (1.61))}$$

$$\therefore \frac{y}{l} = \frac{1}{\beta(1-\beta^2)} \left[\frac{t + \gamma(t)}{\tau_d}\right] - \frac{1}{\beta} \quad (1.65)$$

Taking x-component of the vector Eq. (1.32),

$$-x = \gamma(t) \frac{dx}{dt}$$

$$\therefore \gamma(t) = -\frac{x}{\frac{dx}{dt}} \quad (1.66)$$

1.1. Linear Pursuit

Differentiating Eq. (1.64) w.r.t time,
$$\frac{1}{\tau_d} = -\frac{1}{2}\left\{-(1-\beta^2)\frac{(l-x)^{-\beta}}{l^{1-\beta}} - (1-\beta^2)\frac{(l-x)^{\beta}}{l^{1+\beta}}\right\}\frac{dx}{dt} \quad (1.67)$$

Eq. (1.66) × Eq. (1.67) $\Longrightarrow$
$$\frac{\gamma(t)}{\tau_d} = \frac{x(\beta^2-1)}{2(l-x)}\left\{\left(\frac{l-x}{l}\right)^{1-\beta} + \left(\frac{l-x}{l}\right)^{1+\beta}\right\} \quad (1.68)$$

Eq. (1.64) + Eq. (1.68) $\Longrightarrow$
$$\frac{t+\gamma(t)}{\tau_d}$$
$$= 1 + \frac{1-\beta^2}{2l(l-x)}\left[\frac{(x\beta-l)(l-x)^{1-\beta}}{1-\beta} - \frac{(x\beta+l)(l-x)^{1+\beta}}{1+\beta}\right] \quad (1.69)$$

Eq. (1.69) and Eq. (1.65) $\Longrightarrow$ The tantrik's curve of pursuit is
$$\frac{y}{l} = \frac{1}{\beta(1-\beta^2)} +$$
$$\frac{1}{2l\beta(l-x)}\left[\frac{(x\beta-l)(l-x)^{1-\beta}}{1-\beta} - \frac{(x\beta+l)(l-x)^{1+\beta}}{1+\beta}\right] \quad (1.70)$$
∎

§ Problem 1.1.4. *Assuming the speeds of the tantrik and the ghost from the previous problem be equal in magnitude.*

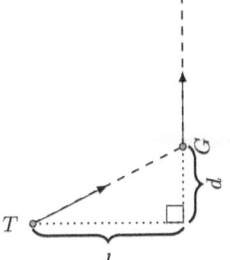

Initial positions of the tantrik T and the ghost G are as shown in the figure.

Find the limit to which the distance TG between the tantrik and the ghost converges.

◇

§§ Solution. Computation of the limit distance is a bit easier in polar co-ordinates relative to the ghost as in Eq. (1.51) and Eq. (1.52).

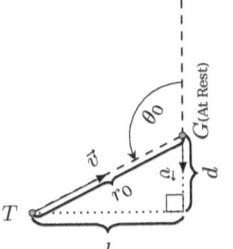

$$\frac{dr}{dt} = -v + v\cos(\pi-\theta)$$
$$\therefore \frac{dr}{dt} = -v(1+\cos\theta)$$
$$(1.71)$$

$$r\frac{d\theta}{dt} = v\sin(\pi-\theta) = v\sin\theta$$
$$(1.72)$$

Now Eq. (1.71) × $(1-\cos\theta)$ + Eq. (1.72) × $\sin\theta$ $\Longrightarrow$
$$\frac{dr}{dt}(1-\cos\theta) + r\frac{d\theta}{dt}\sin\theta = -v(1-\cos^2\theta) + v\sin^2\theta = 0$$

$$\therefore \frac{d}{dt}(r - r\cos\theta) = 0$$
$$r(1 - \cos\theta) = \delta \text{ (a constant)} \qquad (1.73)$$

At $t = 0$, $r_0 = TG = \sqrt{l^2 + d^2}$, $\cos(\pi - \theta_0) = \dfrac{d}{r_0}$.

$$\therefore \sqrt{l^2 + d^2}\left(1 + \frac{d}{\sqrt{l^2 + d^2}}\right) = \delta$$
$$\therefore \delta = d + \sqrt{l^2 + d^2}. \qquad (1.74)$$

In the ghost's frame of reference, it is easy to see that $\theta \to \pi$ when $t \to \infty$.

Also Eq. (1.72) $\implies \dfrac{d\theta}{dt} = \dfrac{v}{r}\sin\theta > 0, \forall \theta < \pi$, $\therefore \theta$ is always increasing till then.

But it is clear from Eq. (1.71) and Eq. (1.72) that $\dfrac{dr}{dt} = 0 = \dfrac{d\theta}{dt}$ when $\theta = \pi$.

Hence, the limiting distance is given by
$$r_{min} = \frac{\delta}{1 - \cos\pi}$$
$$\therefore r_{min} = \frac{d + \sqrt{l^2 + d^2}}{2}. \qquad \blacksquare$$

§ Problem 1.1.5. *A ghost, situated initially at a point G, spell-bound by a tantrik, is confined to move on a fixed straight line GR with a variable speed $v_g(t)$ such that she always stays at a fixed distance l from the tantrik who in turn always moves directly towards her with a fixed speed of v_τ, starting from an initial point T situated at a perpendicular distance of l from GR as shown in the figure.*

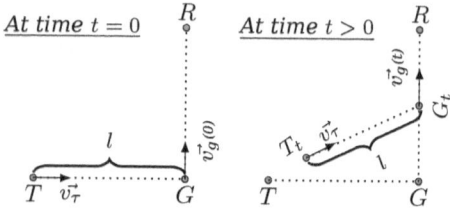

Determine the ghost speed $v_g(t)$ and the tantrik's trajectory. ◇

§§ Solution. Decomposing the tantrik's velocity, relative to the ghost, into radial and transverse components in polar co-ordinates leads to the following equations of motion:

$$r = l = \text{ a constant)}$$
$$\therefore \frac{dr}{dt} = 0$$
$$\frac{dr}{dt} = 0 = -[v_\tau - v_g(t)\cos(\pi - \theta)]$$

$$\therefore v_\tau = -v_g(t)\cos\theta \qquad (1.75)$$

1.1. Linear Pursuit

$$r\frac{d\theta}{dt} = v_g(t)\sin(\pi - \theta)$$

$$\therefore l\frac{d\theta}{dt} = v_g(t)\sin\theta \qquad (1.76)$$

Eq. (1.75) ÷ Eq. (1.76) $\implies$

$$\therefore \frac{v_\tau}{l}dt = -\cot\theta\, d\theta$$

$$\therefore \frac{v_\tau}{l}\int_0^t dt = -\int_{\frac{\pi}{2}}^\theta \cot\theta\, d\theta$$

$$\therefore \frac{v_\tau}{l}t = -\ln\sin\theta$$

$$\therefore \sin\theta = e^{-\frac{v_\tau}{l}t} \qquad (1.77)$$

Hence Eq. (1.75) $\implies$

$$v_g(t) = -\frac{v_\tau}{\cos\theta} = -\frac{v_\tau}{-\sqrt{1-\sin^2\theta}} \quad \because \theta > \frac{\pi}{2}$$

$$\therefore v_g(t) = \frac{v_\tau}{\sqrt{1-e^{-\frac{2v_\tau}{l}t}}}$$

Eq. (1.76) $\implies$

$$\int_0^t v_g(t)\, dt = l\int_{\frac{\pi}{2}}^\theta \operatorname{cosec}\theta\, d\theta$$

$$= -l\ln(\operatorname{cosec}\theta + \cot\theta) = -l\ln\left(\frac{1+\cos\theta}{\sin\theta}\right)$$

$$= -l\ln\left[\frac{1-\cos(\pi-\theta)}{\sin(\pi-\theta)}\right]$$

$$= -l\ln\left[\frac{1-\sqrt{1-\sin^2(\pi-\theta)}}{\sin(\pi-\theta)}\right]$$

$$= -l\ln\left[\frac{1-\sqrt{1-\left(\frac{l-x}{l}\right)^2}}{\frac{l-x}{l}}\right]$$

$$\therefore \int_0^t v_g(t)\, dt = -l\ln\frac{l-\sqrt{x(2l-x)}}{l-x} \qquad (1.78)$$

Also, it is easy to see from the figure that

$$y = \int_0^t v_g(t)\, dt - l\cos(\pi - \theta)$$

$$\therefore -\cos(\pi - \theta) = \frac{y}{l} + \ln\frac{l-\sqrt{x(2l-x)}}{l-x}$$

$$\therefore -\sqrt{1-\sin^2(\pi-\theta)} = \frac{y}{l} + \ln\frac{l-\sqrt{x(2l-x)}}{l-x}$$

$$\therefore -\sqrt{1-\left(\frac{l-x}{l}\right)^2} = \frac{y}{l} + \ln\frac{l-\sqrt{x(2l-x)}}{l-x}$$

Hence the trajectory of the tantrik is

$$y = -\sqrt{x(2l-x)} - l\ln\frac{l-\sqrt{x(2l-x)}}{l-x} \qquad \blacksquare$$

1.1. Linear Pursuit

§ Problem 1.1.6. *A ghost, situated initially at a point G, spell-bound by a tantrik, is confined to move with a constant speed v_g on a fixed straight line GR inclined at an angle δ to TG. R is a safe haven point for the ghost, at and beyond which, she is out of reach of the tantrik.*

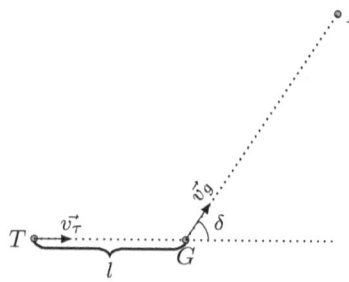

At the same time, the tantrik, situated initially at a point T, such that $TG = l$, always moves directly towards the ghost with a fixed speed of v_τ to catch her.

Compute the distance GR, covered by the ghost, before being caught by the tantrik.

In addition,

  (a) *For a given initial distance between the tantrik and the ghost, $TG = l$, determine the locus of all the ghost's safe haven points which will not result into the capture.*

  (b) *For a given distance moved by the ghost before being caught by the tantrik, $GR = d$, determine the locus of all the tantrik's starting points which will result in capture just before the safe haven point R?*

◊

§§ Solution. Setting up a rectangular co-ordinate system with $T(0,0)$ as the origin, TG as the x-axis, TY as the y-axis and $G(l, 0)$, the computation is as follows.

At some time $t > 0$, let T_t and G_t be the positions of the tantrik and the ghost respectively.

Since, the tantrik is headed always towards the ghost, hence $\vec{v_\tau}$ at T_t is directed towards G_t.

Let γ be the angle between $\vec{v_\tau}$ and TG.

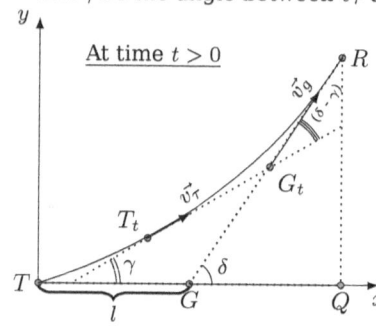

Let τ_d be the time of pursuit.

In τ_d, distance covered by the tantrik along the x-axis is TQ and along the y-axis is QR.

Let $GR = d = v_g \tau_d$.

$$\therefore \tau_d = \frac{d}{v_g} \tag{1.79}$$

$$\therefore TQ = \int_0^{\tau_d} v_\tau \cos\gamma \, dt = TG + GQ = l + d\cos\delta \tag{1.80}$$

1.1. Linear Pursuit

$$\therefore QR = \int_0^{T_d} v_\tau \sin\gamma \, dt = d\sin\delta \tag{1.81}$$

The relative velocity $v_{\tau g}$ of the tantrik with which he is approaching the ghost is given by

$$v_{\tau g} = v_\tau - \vec{v_g} \cdot \frac{\vec{v_\tau}}{v_\tau} = v_\tau - \frac{v_g}{v_\tau}\left[v_\tau \cos(\delta - \gamma)\right]$$

$$\therefore l = \int_0^{T_d} v_{\tau g} \, dt = \int_0^{T_d} v_\tau \, dt - \frac{v_g}{v_\tau}\int_0^{T_d} v_\tau \cos(\delta - \gamma) \, dt$$

$$\therefore l = v_\tau T_d - \frac{v_g}{v_\tau}\left[\cos\delta \int_0^{T_d} v_\tau \cos\gamma \, dt + \sin\delta \int_0^{T_d} v_\tau \sin\gamma \, dt\right]$$

Using Eq. (1.79), Eq. (1.80) and Eq. (1.81)

$$\therefore l = d\frac{v_\tau}{v_g} - \frac{v_g}{v_\tau}\left[\cos\delta\,(l + d\cos\delta) + \sin\delta\,(d\sin\delta)\right]$$

$$= \frac{d}{e} - e\,(d + l\cos\delta), \text{ where } e = \frac{v_g}{v_\tau}.$$

$$\therefore GR = d = \frac{el(1 + e\cos\delta)}{1 - e^2}. \tag{1.82}$$

(a) For a given l, determining the locus of all the ghost's safe haven points, which will not result into the capture, becomes a lot easier with choice of co-ordinate system.

Let the ghost's initial point G be the origin and the line GT be the polar axis.

Hence, as the ghost moves along GR, she makes a polar angle of θ which is equal to the external $\angle TGR$ measured in anti-clockwise direction, starting from GT towards GR. Let $GG_t = r$.

$$\therefore \theta = \pi + \delta$$
$$\therefore \delta = \theta - \pi. \tag{1.83}$$

Putting this value in Eq. (1.82), the required locus is given by

$$r = \frac{el(1 - e\cos\theta)}{1 - e^2}. \tag{1.84}$$

(b) For a given d, determining the locus of all the tantrik's starting points which will result in capture just before the safe haven point R, becomes easier with the following choice of co-ordinate system.

Let the ghost's initial point G be the origin and the line GR be the polar axis.

Initially at T, the tantrik's velocity is directed towards TG. Hence, initially, he makes a polar angle of θ which is equal to the external $\angle RGQ$ measured in anti-clockwise direction, starting from GR towards GQ.

$$\therefore \theta = 2\pi - \delta$$
$$\therefore \delta = 2\pi - \theta. \tag{1.85}$$

Putting this value in Eq. (1.82), the required locus is given by

$$r = \frac{d(1 - e^2)}{e(1 + e\cos\theta)}. \tag{1.86}$$

Note that Eq. (1.86) is an ellipse. ∎

1.2 Cyclic Pursuit

§ Problem 1.2.1. *A ghost, situated initially at a point G, spell-bound by a tantrik, is confined to move with a constant speed v_g on a circular path of radius R in anti-clockwise direction.*

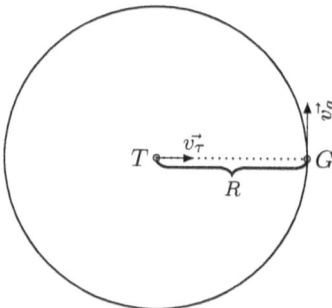

At the same time, the tantrik, situated initially at the center T of the circular path, moves towards the ghost with a fixed speed of v_τ to catch her in such a way that the three points : the center of the circular path, the tantrik and the ghost : are always collinear.

Determine the duration of the chase τ_d, point of intersection (if caught), distance covered by the ghost and the tantrik in τ_d and equation of the tantrik's trajectory. ◊

§§ Solution. Assuming polar co-ordinates with T as the pole and TG as the polar axis, at some time $t > 0$, let $G_t(R, \theta)$ and $T_t(r, \theta)$ be the positions of the ghost and the tantrik respectively.

<u>At Time $t > 0$</u>

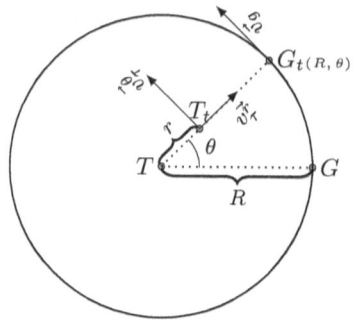

It is given that the points T, T_t and G_t are collinear, hence TT_tG_t is a straight line.
Let v_g^r be the radial component and v_g^θ be the transverse components of the ghost's velocity v_g respectively. Since the ghost moves always on the circle,

$$\therefore v_g^r = 0, \ v_g^\theta = R\frac{d\theta}{dt}.$$

$$\therefore \vec{v_g} = v_g^r \ \hat{\mathbf{r}} + v_g^\theta \ \hat{\theta} = R\frac{d\theta}{dt}\hat{\theta}$$

$$\therefore v_g = |\vec{v_g}| = R\frac{d\theta}{dt} \quad (1.87)$$

Let v_τ^r be the radial component and v_τ^θ be the transverse components of the tantrik's velocity v_τ respectively.

$$\therefore v_\tau^r = \frac{dr}{dt}, \ v_\tau^\theta = r\frac{d\theta}{dt} = \frac{r}{R}v_g \text{ (Using Eq. (1.87))}$$

$$\therefore \vec{v_\tau} = v_\tau^r \ \hat{\mathbf{r}} + v_\tau^\theta \ \hat{\theta} = \frac{dr}{dt} \ \hat{\mathbf{r}} + \frac{r}{R}v_g \ \hat{\theta}$$

$$\therefore v_\tau^2 = \left(\frac{dr}{dt}\right)^2 + \frac{r^2}{R^2}v_g^2$$

$$\therefore \left(\frac{dr}{dt}\right)^2 = \frac{v_g^2}{R^2}\left[\left(\frac{v_\tau}{v_g}R\right)^2 - r^2\right]$$

$$\therefore \frac{dr}{\sqrt{\left(\frac{v_\tau}{v_g}R\right)^2 - r^2}} = \frac{v_g}{R} \ dt$$

1.2. Cyclic Pursuit

$$\therefore \int_0^R \frac{dr}{\sqrt{\left(\frac{v_\tau}{v_g}R\right)^2 - r^2}} = \frac{v_g}{R}\int_0^{\tau_d} dt$$

$$\therefore \sin^{-1}\left(\frac{r}{\frac{v_\tau}{v_g}R}\right)\bigg|_{r=0}^{r=R} = \frac{v_g}{R}\tau_d$$

$$\therefore \tau_d = \frac{R}{v_g}\sin^{-1}\frac{v_g}{v_\tau}. \qquad (1.88)$$

Distance covered by the ghost in τ_d is $v_g\tau_d$, which is

$$R\sin^{-1}\frac{v_g}{v_\tau}.$$

Distance covered by the tantrik in τ_d is $v_\tau\tau_d$, which is

$$\frac{v_\tau}{v_g}R\sin^{-1}\frac{v_g}{v_\tau}.$$

Assuming θ_τ to be angle traversed by the ghost in τ_d, integrating Eq. (1.87),

$$\int_0^{\theta_\tau} d\theta = \frac{v_g}{R}\int_0^{\tau_d} dt$$

$$\therefore \theta_\tau = \frac{v_g\tau_d}{R} = \sin^{-1}\frac{v_g}{v_\tau}.$$

Hence, the point of intersection is

$$(R, \theta_\tau) \equiv \left(R,\ \sin^{-1}\frac{v_g}{v_\tau}\right).$$

Note that the equation leading to Eq. (1.88) describes the tantrik's trajectory, which is

$$\sin^{-1}\left(\frac{r(t)}{\frac{v_\tau}{v_g}R}\right) = \frac{v_g}{R}t = \theta(t) \because \frac{d\theta}{dt} = \frac{v_g}{R}$$

$$\therefore r(t) = \mu R \sin\theta(t),\ (\text{where } \mu = \frac{v_\tau}{v_g}) \qquad (1.89)$$

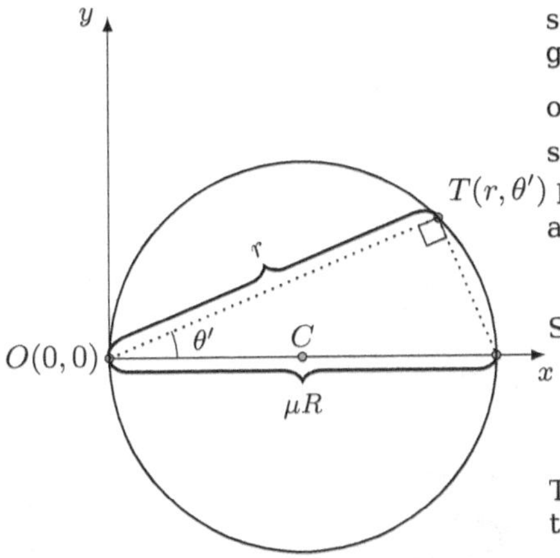

It is easy to see that this describes a circle with the origin $C\left(\frac{\mu R}{2}, 0\right)$ and the radius of length OC as $\frac{\mu R}{2}$, as described ahead.

$T(r, \theta')$ Eq. (1.89) can be expressed as

$$r(t) = \mu R \cos\left[\frac{\pi}{2} - \theta(t)\right]$$

Substituting $\theta' = \frac{\pi}{2} - \theta(t)$

$$r(t) = \mu R \cos\theta'(t)$$

$$\therefore \cos\theta'(t) = \frac{r(t)}{\mu R}$$

This is a circle as drawn in the figure before.

∎

1.2. Cyclic Pursuit

§ Problem 1.2.2. *A ghost, situated initially at a point G, spell-bound by a tantrik, is confined to move with a constant speed v_g on a circular path of radius R in anti-clockwise direction.*

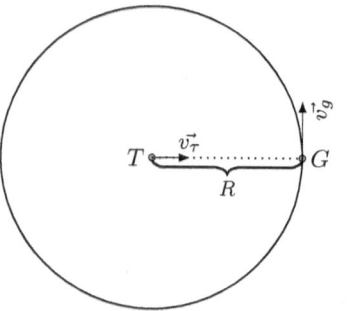

At the same time, the tantrik, situated initially at the center T of the circular path, always moves directly towards the ghost with a fixed speed of v_τ to catch her.

Determine the equations of the curve of pursuit. ◊

§§ Solution. It is given that the velocity vector of the tantrik is always pointed towards the instantaneous position of the ghost.

Computation of the tantrik's curve of pursuit will be done relative to the ghost as follows.

<u>At Time $t > 0$</u>

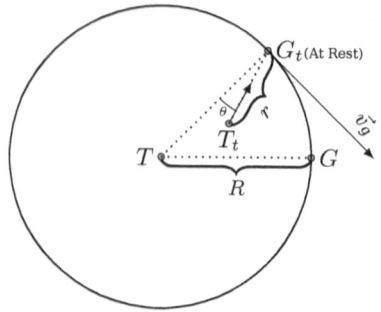

At some time $t > 0$, let G_t and T_t be the positions of the ghost and the tantrik respectively.

Assuming polar coordinates with the point G_t as the pole and $G_t T$ as the polar axis, $G_t T_t = r$, $\angle TG_t T_t = \theta$.

Resolving the tantrik's velocity in radial and transverse directions respectively:

$$\frac{dr}{dt} = -\left[v_\tau - v_g \cos\left(\frac{\pi}{2} - \theta\right)\right]$$

$$\therefore \frac{dr}{dt} = v_g \sin\theta - v_\tau \quad (1.90)$$

And $\quad r\dfrac{d\theta}{dt} = v_g \sin\left(\dfrac{\pi}{2} - \theta\right)$

$$\therefore r\frac{d\theta}{dt} = v_g \cos\theta \quad (1.91)$$

Eq. (1.90) $\div$ Eq. (1.91) $\implies$

$$\frac{dr}{r} = (\tan\theta - \mu\sec\theta)\,d\theta, \text{ (where } \mu = \frac{v_\tau}{v_g}\text{)}$$

$$\therefore \int_R^r \frac{dr}{r} = \int_0^\theta (\tan\theta - \mu\sec\theta)\,d\theta$$

$$\therefore \ln \frac{r}{R} = -\ln\cos\theta - \ln(\sec\theta + \tan\theta) = \ln\frac{\sec\theta}{(\sec\theta + \tan\theta)^\mu}$$

$$\therefore r = \frac{R\sec\theta}{(\sec\theta + \tan\theta)^\mu} \qquad \blacksquare$$

1.3 Triangular Pursuit

§ Problem 1.3.1. *Three ghosts, G_1, G_2 and G_3, situated initially at the three vertices of an equilateral triangle $G_1G_2G_3$, spell-bound by a tantrik T who in turn is situated at the centroid of the triangle, are confined to move with a constant speed v_g towards their nearest neighbour in anti-clockwise direction.*

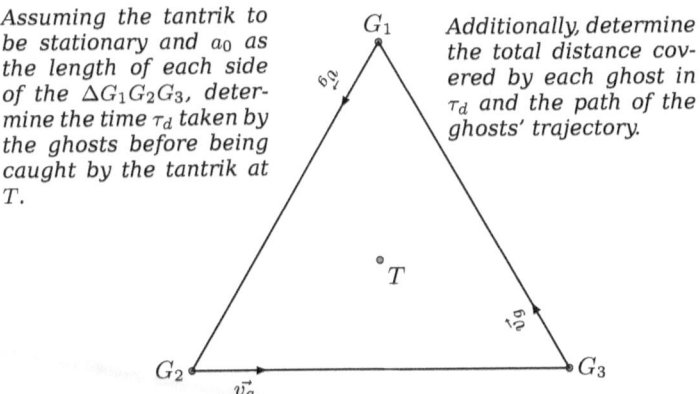

Assuming the tantrik to be stationary and a_0 as the length of each side of the $\Delta G_1G_2G_3$, determine the time τ_d taken by the ghosts before being caught by the tantrik at T.

Additionally, determine the total distance covered by each ghost in τ_d and the path of the ghosts' trajectory.

Also, determine the number of revolutions executed by each ghost in a given time $\eta\tau_d$, where $0 < \eta < 1$. ◊

§§ Solution.

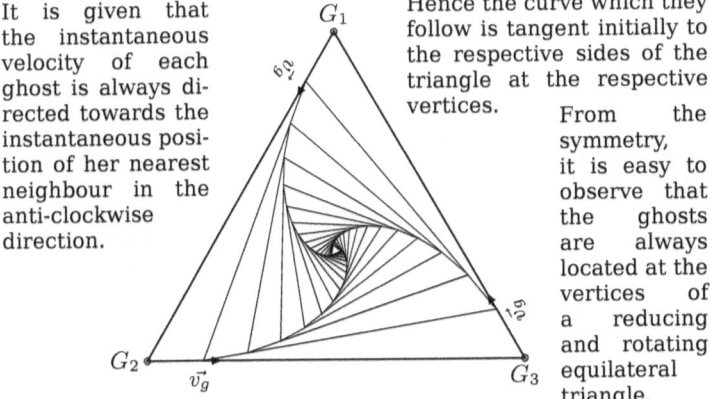

It is given that the instantaneous velocity of each ghost is always directed towards the instantaneous position of her nearest neighbour in the anti-clockwise direction.

Hence the curve which they follow is tangent initially to the respective sides of the triangle at the respective vertices. From the symmetry, it is easy to observe that the ghosts are always located at the vertices of a reducing and rotating equilateral triangle.

Because the speed of each ghost is the same, hence the vertices of the respective triangles are always equidistant from the centroid T of the initial triangle $G_1G_2G_3$, thus T is the centroid of the respective triangles too.

1.3. Triangular Pursuit

Let T be the origin for a system of plane polar co-ordinates.

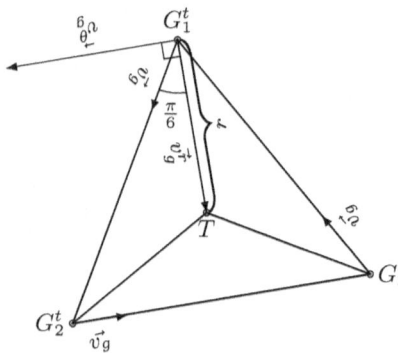

Let G_1^t, G_2^t and G_3^t be the respective positions of the ghosts at some time $t > 0$, such that $TG_1^t = TG_2^t = TG_3^t = r$.

Due to the symmetry, the ghosts get caught by the tantrik at T simultaneously.

The ghost G_1 is at a distance r from the origin T and is moving with a velocity $\vec{v_g}$ towards the ghost G_2.

Resolving $\vec{v_g}$ into radial and transverse components respectively:
$$\vec{v_g} = v_g^r\, \hat{\mathbf{r}} + v_g^\theta\, \hat{\theta}$$

$$v_g^r = \frac{dr}{dt} = -v_g \cos \frac{\pi}{6} = -\frac{v_g \sqrt{3}}{2} \tag{1.92}$$

$$v_g^\theta = r\frac{d\theta}{dt} = v_g \sin \frac{\pi}{6} = \frac{v_g}{2} \tag{1.93}$$

Eq. (1.92) $\div$ Eq. (1.93) $\implies$

$$\frac{dr}{r} = -\sqrt{3} d\theta$$

$$\therefore \int_{r_0}^{r} \frac{dr}{r} = -\sqrt{3} \int_0^\theta d\theta$$

$$\ln \frac{r}{r_0} = -\theta \sqrt{3}$$

$$\therefore r = r_0 e^{-\theta \sqrt{3}} \tag{1.94}$$

Here r_0 is the initial value of r, i.e., the initial distance of the ghost from the tantrik T. It is easy to see that $r_0 = \dfrac{a_0}{2\cos \dfrac{\pi}{6}} = \dfrac{a_0}{\sqrt{3}}$.

$$\therefore r = \frac{a_0}{\sqrt{3}} e^{-\theta \sqrt{3}} \tag{1.95}$$

The pursuit curve of each ghost is thus an equiangular logarithmic spiral with its pole at the centroid T of the triangle and tangent at each vertex of the triangle to one of the sides.

Note that Eq. (1.95) is independent of the ghost's speed v_g, hence the pursuit curve is the same even for non-constant speed.

Distance S traveled by each ghost in time τ_d is given by
$$S = \int_0^{\tau_d} v_g dt = -\frac{2}{\sqrt{3}} \int_{r_0}^{0} dr \text{ (Using Eq. (1.92))}$$

$$\therefore S = \frac{2}{\sqrt{3}} r_0 = \frac{2}{\sqrt{3}} \times \frac{a_0}{\sqrt{3}}$$

$$\therefore S = \frac{2}{3} a_0 \tag{1.96}$$

To compute the duration τ_d of the pursuit, integrating Eq. (1.92)

1.3. Triangular Pursuit

gives:
$$\int_{r_0}^{0} dr = -\frac{v_g\sqrt{3}}{2} \int_0^{\tau_d} dt$$
$$\therefore \tau_d = \frac{2}{v_g\sqrt{3}} r_0 = \frac{2}{v_g\sqrt{3}} \times \frac{a_0}{\sqrt{3}}$$
$$\therefore \tau_d = \frac{2a_0}{3v_g} \tag{1.97}$$

Note that $\tau_d = \dfrac{S}{v_g}$.

Alternatively, the ghost G_1 travels the radial distance $S_r = G_1 T = r_0 = \dfrac{a_0}{\sqrt{3}}$ with the radial component of her velocity $v_g^r = \dfrac{v_g\sqrt{3}}{2}$.

Hence the duration of pursuit $\tau_d = \dfrac{r_0}{v_g^r} = \dfrac{2a_0}{3v_g}$, which is the same as Eq. (1.97).

And the distance traveled by the ghost in time τ_d is $S = v_g \tau_d = \dfrac{2}{3}a_0$, which is the same as Eq. (1.96).

Note that the transverse distance traveled by each ghost in time τ_d is given by $S_\theta = v_g^\theta \tau_d = \dfrac{v_g}{2} \times \dfrac{2a_0}{3v_g} = \dfrac{a_0}{3}$.

And it is easy to verify these results with
$$S = \sqrt{S_r^2 + S_\theta^2} = \sqrt{\frac{a_0^2}{3} + \frac{a_0^2}{9}} = \frac{2}{3}a_0,$$
which is the same as Eq. (1.96).

The last computation hints at another way (utilizing polar coordinates) to compute the total distance S traveled by each ghost, which is as follows.
$$S = \int ds = \int \sqrt{(dr)^2 + (rd\theta)^2}$$
$$= \int_0^\infty d\theta \sqrt{\left(\frac{dr}{d\theta}\right)^2 + r^2} \quad \because \text{ as } \theta \to \infty,\ r \to 0$$

Eq. (1.92) ÷ Eq. (1.93) $\implies \dfrac{dr}{d\theta} = -r\sqrt{3}$

$$\therefore S = \int_0^\infty d\theta \sqrt{3r^2 + r^2} = 2\int_0^\infty r\, d\theta$$
$$= 2\int_0^\infty \frac{a_0}{\sqrt{3}} e^{-\theta\sqrt{3}} d\theta \text{ (Using Eq. (1.95))}$$
$$= \frac{2a_0}{\sqrt{3}} \int_0^\infty e^{-\theta\sqrt{3}} d\theta = \frac{2a_0}{\sqrt{3}} \left(\frac{e^{-\theta\sqrt{3}}}{-\sqrt{3}}\right)\bigg|_0^\infty$$
$$\therefore S = \frac{2}{3}a_0$$

which is the same as Eq. (1.96).

To determine the number of revolutions made by each ghost, computing $\theta(t)$ is a prerequisite.

1.3. Triangular Pursuit

Integrating Eq. (1.92) gives
$$\int_{r_0}^{r} dr = -\frac{v_g\sqrt{3}}{2}\int_0^t dt, \text{ (where } r_0 = \frac{a_0}{\sqrt{3}}\text{)}$$
$$\therefore r = \frac{a_0}{\sqrt{3}}\left(1 - \frac{3v_g}{2a_0}t\right)$$

Using Eq. (1.97):
$$\therefore r = \frac{a_0}{\sqrt{3}}\left(1 - \frac{t}{\tau_d}\right) \tag{1.98}$$

Integrating Eq. (1.93) gives
$$\int_0^\theta d\theta = \frac{v_g}{2}\int_0^t \frac{1}{r} dt$$
$$\therefore \theta = \frac{v_g\sqrt{3}}{2a_0}\int_0^t \frac{1}{\left(1 - \frac{t}{\tau_d}\right)} dt$$

Let $u = 1 - \frac{t}{\tau_d}$. Then $du = -\frac{1}{\tau_d} dt$.
$$\therefore \theta = -\frac{\tau_d v_g\sqrt{3}}{2a_0}\int_1^u \frac{1}{u} du$$
$$= -\frac{\tau_d v_g\sqrt{3}}{2a_0}\ln u = -\frac{\tau_d v_g\sqrt{3}}{2a_0}\ln\left(1 - \frac{t}{\tau_d}\right)$$
$$\therefore \theta = -\frac{1}{\sqrt{3}}\ln\left(1 - \frac{t}{\tau_d}\right) \tag{1.99}$$

Putting $t = \eta\tau_d$ in Eq. (1.99):
$$\theta(t = \eta\tau_d) = -\frac{1}{\sqrt{3}}\ln(1 - \eta) = \frac{1}{\sqrt{3}}\ln\frac{1}{1 - \eta}$$

Hence the number of revolutions executed by each ghost in a given time $\eta\tau_d$ is
$$\frac{\theta(t = \eta\tau_d)}{2\pi} = \frac{1}{2\pi\sqrt{3}}\ln\frac{1}{1 - \eta}.$$

∎

Chapter 2

Motorboat Problem and Crossing The River

2.1 Rendezvous with Floating Raft

§ Problem 2.1.1. *A motorboat going upstream overcame a raft floating downstream. τ later the engine of the boat was stopped for τ_α. Then the boat traveled downstream and passed the raft in next τ_β. The speed of the boat relative to the water was constant and equal to v_b. Determine the river current velocity v_0, considering it constant.* ◊

§§ Solution. *Graphical Approach*: It is easier to solve the problem in a coordinate system related to the water. The speed of the raft equal to the velocity of the river current is zero in this system, and the speed of the boat upstream (OA) and downstream (BC) will be the same in magnitude, hence $\tan\theta = v_b$ as in Distance-Time graph of the boat:

In $\triangle$ ADO:
$$\tan\theta = \frac{AD}{OD} = \frac{AD}{\tau}$$
$$\therefore AD = PE = \tau\tan\theta = v_b\tau$$

In $\triangle$ BEC:
$$\tan\theta = \frac{BE}{EC} = \frac{BE}{\tau_\beta}$$
$$\therefore BE = \tau_\beta\tan\theta = v_b\tau_\beta$$

When the boat-engine was stopped (AB), the boat's speed with respect to the water was equal to the river current velocity v_0. Hence $\tan\phi = v_0$.

2.1. Rendezvous with Floating Raft

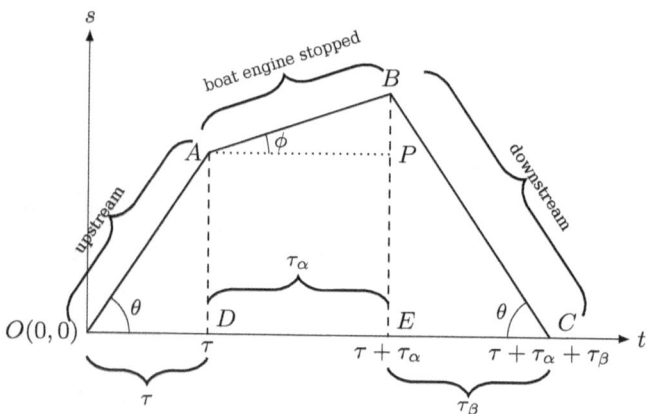

In △ APB:
$$\tan\phi = \frac{BP}{AP} = \frac{BE - PE}{DE} = \frac{v_b\tau_\beta - v_b\tau}{\tau_\alpha}$$
$$\therefore v_0 = \tan\phi = v_b\frac{\tau_\beta - \tau}{\tau_\alpha}.$$

Analytic Approach: From the moment the boat overcame the raft to the moment it passed it again, the raft covered a distance equal to $s = v_0(\tau + \tau_\alpha + \tau_\beta)$.

On the other hand, this distance is equal to the difference between the distances traveled by the boat upstream and downstream because when the engine was stopped, the distance between the boat and the raft did not increase: $s = \tau_\beta(v_b + v_0) - \tau(v_b - v_0)$.
$$\therefore v_0(\tau + \tau_\alpha + \tau_\beta) = \tau_\beta(v_b + v_0) - \tau(v_b - v_0)$$
$$\therefore v_0 = v_b\frac{\tau_\beta - \tau}{\tau_\alpha}. \qquad \blacksquare$$

§ Problem 2.1.2. *A motorboat going upstream overcame a raft floating downstream. τ after this the engine of the boat stalled. It took τ_α to repair it, and during this time the boat freely floated downstream. When the engine was repaired, the boat traveled downstream with the same speed relative to the current as before and passed the raft at a distance of l from the point where they had met the first time. Determine the velocity of the river current, considering it constant.* ◊

§§ Solution. During τ after overcoming the raft, the boat traveled away from it. During the next τ_α, when the engine was being repaired, the distance between the boat and the raft did not increase. The boat overtook the raft in τ because its speed with respect to the water and hence to the raft was constant.

Thus the velocity of the current
$$v_0 = \frac{l}{t} = \frac{l}{\tau + \tau_\alpha + \tau} = \frac{l}{2\tau + \tau_\alpha}. \qquad \blacksquare$$

2.1. Rendezvous with Floating Raft

§ Problem 2.1.3. *A motorboat going downstream overcame a raft at a point A; τ later it turned back and after some time passed the raft at a distance l from the point A. Find the flow velocity. Assuming duty of the engine to be constant.*   ◊

§§ Solution. Since duty of the engine was constant, hence the speed of the boat relative to the water was constant and equal to v_b (say). After turning back, the boat passed the raft after τ_β (say).

Graphical Approach: It is easier to solve the problem in a co-ordinate system related to the water. The speed of the raft equal to the velocity of the river current is zero in this system, and the speed of the boat downstream (AB) and upstream (BC) will be the same in magnitude, hence $\tan\theta = v_b$ as in Distance-Time graph of the boat:

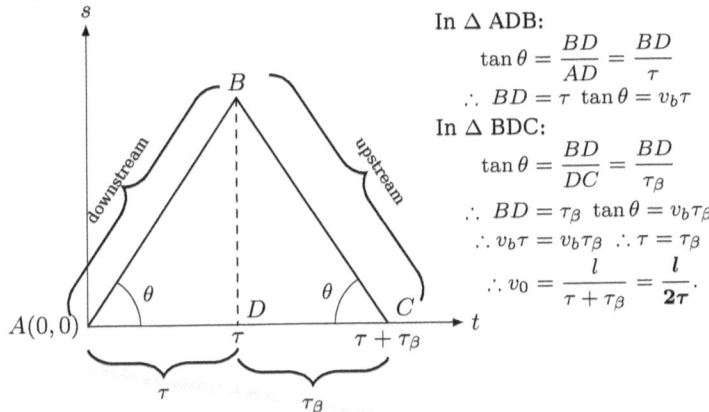

In $\triangle$ ADB:
$$\tan\theta = \frac{BD}{AD} = \frac{BD}{\tau}$$
$$\therefore BD = \tau\tan\theta = v_b\tau$$

In $\triangle$ BDC:
$$\tan\theta = \frac{BD}{DC} = \frac{BD}{\tau_\beta}$$
$$\therefore BD = \tau_\beta\tan\theta = v_b\tau_\beta$$
$$\therefore v_b\tau = v_b\tau_\beta \quad \therefore \tau = \tau_\beta$$
$$\therefore v_0 = \frac{l}{\tau+\tau_\beta} = \frac{l}{2\tau}.$$

Analytic Approach: From the moment the boat overcame the raft to the moment it passed it again, the raft covered a distance equal to $l = v_0(\tau+\tau_\beta)$.

On the other hand, this distance is equal to the difference between the distances traveled by the boat upstream and downstream: $l = \tau_\beta(v_b+v_0) - \tau(v_b-v_0)$.

$$\therefore v_0(\tau+\tau_\beta) = \tau_\beta(v_b+v_0) - \tau(v_b-v_0)$$
$$\therefore \tau_\beta = \tau$$
$$\therefore l = v_0(\tau+\tau_\beta) = v_0(2\tau)$$
$$\therefore v_0 = \frac{l}{2\tau}.$$

Alternatively: During τ after overcoming the raft, the boat traveled away from it. The boat overtook the raft in τ because its speed with respect to the water and hence to the raft was constant. Thus the velocity of the current $v_0 = \frac{l}{t} = \frac{l}{\tau+\tau} = \frac{l}{2\tau}.$ ∎

§ Problem 2.1.4. *A boatman lowers a wooden bucket into the water and himself sets off downstream, rowing. After τ_a he teaches a point A, l from his starting-point and turns back. He picks up the buck-et,*

2.1. Rendezvous with Floating Raft

turns round again and, rowing downstream once more, reaches A for the second time τ_b later. Assuming that the speeds of both current and boat are constant, and also that no time is wasted on turning round?

(a) How long does the oarsman spend on rowing upstream to meet the bucket?
(b) Determine the meeting point of the boat and bucket in second round.
(c) What is the speed of the current?
(d) What is the boat's speed relative to the water? ◊

§§ Solution. Relative to the bucket (which floats downstream) the boat's speed upstream and downstream must be the same.

(a) Thus the boatman also spends τ_a on returning to meet the bucket.

(b) Further, the speed of the boat downstream relative to the bank is $\dfrac{l}{\tau_a}$ (since it takes τ_a to travel l to point A),

$$\therefore v_b + v_0 = \frac{l}{\tau_a},$$ where v_b is boat's speed relative to the water and v_0 is speed of the current.

After meeting the bucket, the boat spends τ_b on returning downstream,

so the meeting place is $(v_b+v_0)\tau_b = l\dfrac{\tau_b}{\tau_a}$ above A, or $l\left(1 - \dfrac{\tau_b}{\tau_a}\right)$ from the place where the bucket is lowered into the water.

(c) Thus the bucket travels $l\left(1 - \dfrac{\tau_b}{\tau_a}\right)$ in $2\tau_a$ between being lowered and picked up, i.e, the speed of the current

$$v_0 = \frac{l\left(1 - \dfrac{\tau_b}{\tau_a}\right)}{2\tau_a} = \frac{l}{2\tau_a}\left(1 - \frac{\tau_b}{\tau_a}\right).$$

(d) The speed of the boat relative to the water is
$$v_b = (v_b + v_0) - v_0$$
$$= \frac{l}{\tau_a} - \frac{l}{2\tau_a}\left(1 - \frac{\tau_b}{\tau_a}\right)$$
$$= \frac{l}{2\tau_a}\left(1 + \frac{\tau_b}{\tau_a}\right).$$ ∎

§ Problem 2.1.5. *Two motorboats were going downstream with different velocities v_α and v_β respectively. When one overtook the other a ring-buoy was dropped from one of the boats. τ time later both boats turned back simultaneously and went at the same speeds as before (relative to the water) to the spot where the ring-buoy had been dropped.*

Which of the launches will reach the ring-buoy first?
Solve this problem also for the cases in which the boats:

(a) *went upstream; and*

(b) *were approaching each other*

before they met. ◊

2.2. Rendezvous with Anchored Ports

§§ Solution. The velocity of the river current v_0 affects the motion of both boats and the ring-buoy similarly and cannot change their mutual positions. This velocity may therefore be disregarded and the motion of the boats and the ring-buoy only considered relative to the water.

The distances traversed by the boats relative to the water in the time τ before they turn will be
$$d_\alpha = v_\alpha \, \tau$$
and
$$d_\beta = v_\beta \, \tau.$$

Returning to the ring-buoy with the original velocities v_α and v_β the boats should obviously spend the same time to cover the distances d_α and d_β to the ring-buoy as was spent when they steamed away from it.

Hence in all the three cases the motorboats will meet the ring-buoy at the same time. ∎

§ Problem 2.1.6. *The speed of a motorboat with respect to the water is v_b, the speed of the stream v_0. When the boat began traveling upstream, a float was dropped from it. The boat traveled a distance l upstream, turned about and caught up with the float. How long is it before the boat reaches the float again?* ◊

§§ Solution. *In a reference frame fixed relative to the bank* : Moving upstream at a speed $v_b - v_0$ the boat covered a distance l. Moving downstream at a speed $v_b + v_0$ it covered a greater distance, namely $l + v_0 \tau$, where $v_0 \tau$ is the distance covered by the float with respect to the bank.

The time of motion is
$$\tau = \frac{l}{v_b - v_0} + \frac{l + v_0 \, \tau}{v_b + v_0}.$$
$$\therefore \tau = \frac{2l}{v_b - v_0}.$$

In a reference frame fixed with respect to the water : The speed of the boat in this reference frame is v_b, the float is at rest. The time of motion of the boat is $\tau = \dfrac{2l_b}{v_b}$, where l_b is the distance covered by the boat with respect to the water in one direction. It is easy to see that
$$\frac{l}{v_b - v_0} = \frac{l_b}{v_b} = \frac{\tau}{2},$$
$$\therefore \tau = \frac{2l}{v_b - v_0}.$$

Hence it is easy to solve in a moving reference frame. ∎

2.2 Rendezvous with Anchored Ports

§ Problem 2.2.1. *Two boats, leave the two buoys anchored at the positions A and B respectively in a river, meet each other and return. If the boats leave their respective buoys at the same time, the boat*

2.2. Rendezvous with Anchored Ports

departing from A spends τ_a both ways and that from B, τ_b. The speed of both boats are the same relative to the water.
(a) *Determine how much later should the boat depart from A after the other one leaves B for the two to travel the same time. Determine the point where the boats will meet and the meeting time.*
(b) *Determine the speed of the boats with respect to the water, the velocity of the river current, the point where the boats will meet and the meeting time if they leave their respective buoys simultaneously. The distance between the buoys is l.* ◊

§§ Solution. The motion of the boats leaving their buoys at the same time is shown by lines AC_1A_n and BC_1B_1, where C_1 is their meeting point as depicted in the distance-time graph of the boats. Since the speeds of the boats relative to the water are the same, AB_1 and BA_n are straight lines.

(a) The motion of the boat leaving the buoy B at the designated time is shown by line BC_2B_2 and that of the boat leaving the buoy A at the time τ_0 later is shown by line $A_2C_2A_n$, where C_2 is their meeting point. Since the speeds of the boats relative to the water are the same, A_2B_2 and BA_n are straight lines, the points B, C_1, C_2 are collinear and $AB_1 \parallel A_2B_2$.

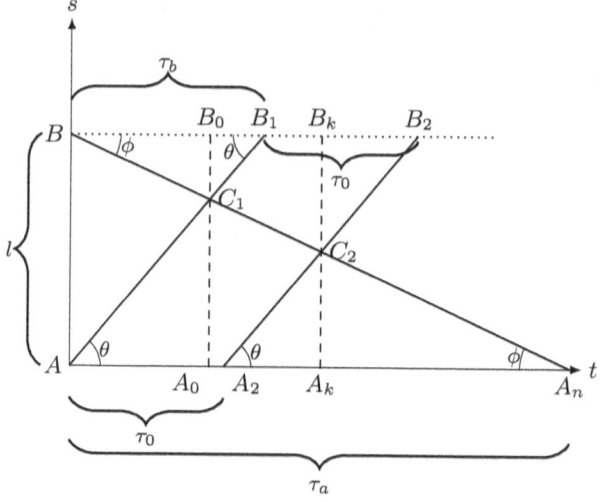

Since both travel the same time,
$$\therefore BB_2 = A_2A_n \quad \therefore BB_1 + B_1B_2 = AA_n - AA_2.$$
$$\therefore \tau_b + \tau_0 = \tau_a - \tau_0$$
$$\therefore \tau_0 = \frac{\tau_a - \tau_b}{2}.$$

It can been that $\triangle BC_2B_2$ and $\triangle A_2C_2A_n$ are similar.
$$\therefore \frac{A_2A_n}{BB_2} = \frac{A_2C_2}{C_2B_2} = \frac{A_nC_2}{BC_2} = \frac{A_kC_2}{C_2B_k} = 1.$$
$\therefore C_2$ is the mid-point of the straight lines A_2B_2, BA_n and $A_kB_k = AB$. $\therefore A_kC_2 = \dfrac{l}{2}$.

2.2. Rendezvous with Anchored Ports

Hence the point of meeting is at $\frac{l}{2}$.

In $\triangle A_2A_kC_2$:
$$\tan\theta = \frac{A_kC_2}{A_2A_k} = \frac{l}{2A_2A_k} = \frac{l}{\tau_b} \text{ from } \triangle ABB_1.$$
$$\therefore A_2A_k = \frac{\tau_b}{2}.$$

Hence meeting time =
$$AA_k = AA_2 + A_2A_k = \tau_0 + \frac{\tau_b}{2} = \frac{\tau_a - \tau_b}{2} + \frac{\tau_b}{2} = \boldsymbol{\frac{\tau_a}{2}}.$$

(b) The speed of the boats with respect to the water v_b and the velocity of the river current v_0 can be found from the slopes in downstream and upstream paths of motion respectively
$$\because \tan\theta = v_b + v_0 \quad \& \quad \tan\phi = v_b - v_0.$$

But in $\triangle ABB_1$:
$$\tan\theta = \frac{AB}{BB_1} = \frac{l}{\tau_b}$$

In $\triangle BAA_n$:
$$\tan\phi = \frac{AB}{AA_n} = \frac{l}{\tau_a}.$$

$$\therefore v_b + v_0 = \frac{l}{\tau_b} \quad \& \quad v_b - v_0 = \frac{l}{\tau_a}.$$
$$\therefore v_b = \frac{l(\tau_a + \tau_b)}{2\tau_a\tau_b} \quad \& \quad v_0 = \frac{l(\tau_a - \tau_b)}{2\tau_a\tau_b}.$$

Alternatively, these can be found from the equations $l = \tau_b(v_b + v_0)$ and $l = \tau_a(v_b - v_0)$, where τ_a and τ_b are the times of motion of the boats downstream and upstream.

In $\triangle AA_0C_1$: $\tan\theta = \dfrac{A_0C_1}{AA_0} = \dfrac{l}{\tau_b}$ from $\triangle ABB_1$.

In $\triangle BB_0C_1$: $\tan\phi = \dfrac{C_1B_0}{BB_0} = \dfrac{l}{\tau_a}$ from $\triangle BAA_n$.

$\because AA_0 = BB_0$, adding these:
$$A_0C_1 + C_1B_0 = l\left(\frac{1}{\tau_a} + \frac{1}{\tau_b}\right) \quad AA_0 = A_0B_0 = AB = l$$
$$\therefore AA_0 = \frac{\tau_a\tau_b}{\tau_a + \tau_b}$$
$$\therefore A_0C_1 = \frac{l}{\tau_b}AA_0 = l\frac{\tau_a}{\tau_a + \tau_b}.$$

Hence the point of the meeting is at a distance of $l\dfrac{\tau_a}{\tau_a + \tau_b}$ **from the buoy** A **at time** $\dfrac{\tau_a\tau_b}{\tau_a + \tau_b}$.

Alternatively,
$$l = (v_b + v_0)\tau_m + (v_b - v_0)\tau_m$$
$$= \tau_b(v_b + v_0)$$
$$= \tau_a(v_b - v_0)$$
where τ_m = meeting time.
$$\therefore \tau_m = \frac{l}{2v_b} = \frac{\tau_a\tau_b}{\tau_a + \tau_b}.$$

2.2. Rendezvous with Anchored Ports

And distance of the meeting point from the buoy $A =$
$$(v_b + v_0)\tau_m = l\frac{\tau_a}{\tau_a + \tau_b}.$$

Alternatively,
It can been that $\triangle BC_1 B_1$ and $\triangle AC_1 A_n$ are similar.
$$\therefore \frac{AA_n}{BB_1} = \frac{A_0 C_1}{C_1 B_0}$$
$$\therefore \frac{A_0 C_1}{\tau_a} = \frac{C_1 B_0}{\tau_b} = \frac{A_0 C_1 + C_1 B_0}{\tau_a + \tau_b} = \frac{l}{\tau_a + \tau_b}$$
$$\therefore A_0 C_1 = l\frac{\tau_a}{\tau_a + \tau_b}. \qquad \blacksquare$$

§ Problem 2.2.2. *Determine the minimum distance between two motorboats simultaneously leaving from two ports A and B separated by a distance l, with uniform velocities v_a and v_b respectively. The paths of motion of the boats is at angles α and β with AB respectively.* ◊

§§ Solution. Assuming the origin of the coordinate-system at the port A, the equations of motion of the boats are

$$\mathbf{r_a} = v_a t \,\cos\alpha\,\hat{\mathbf{i}} + v_a t \,\sin\alpha\,\hat{\mathbf{j}}$$
$$\mathbf{r_b} = (l - v_b t \,\cos\beta)\,\hat{\mathbf{i}} + v_b t \,\sin\beta\,\hat{\mathbf{j}}$$

Hence distance between the boats,
$$r = |\mathbf{r_b} - \mathbf{r_a}|$$
$$= \sqrt{[l - (v_b \cos\beta + v_a \cos\alpha)\,t]^2 + [(v_b \sin\beta - v_a \sin\alpha)\,t]^2}$$

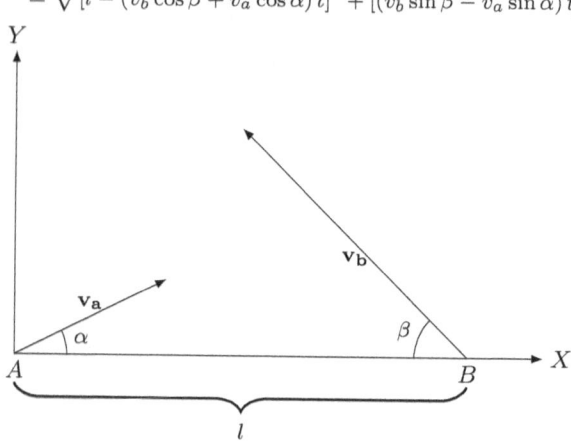

$$\therefore r^2 = [l - (v_b \cos\beta + v_a \cos\alpha)\,t]^2$$
$$+ [(v_b \sin\beta - v_a \sin\alpha)\,t]^2$$

Differentiating both sides w.r.t. t
$$2r\dot{r} = -2\,[l - (v_b \cos\beta + v_a \cos\alpha)\,t]\,(v_b \cos\beta + v_a \cos\alpha)$$
$$+ 2\,(v_b \sin\beta - v_a \sin\alpha)^2\,t.$$

2.2. Rendezvous with Anchored Ports

Equating it to zero to minimize the expression : we exclude the trivial case where $r = 0$; which would mean that the boats would eventually collide, hence $\dot{r} = 0 \implies$

$$t = \frac{l(v_a \cos\alpha + v_b \cos\beta)}{v_a^2 + v_b^2 + 2v_a v_b \cos(\alpha+\beta)};$$

$$\therefore r_{min} = \frac{l(v_b \sin\beta - v_a \sin\alpha)}{\sqrt{v_a^2 + v_b^2 + 2v_a v_b \cos(\alpha+\beta)}}.$$

Alternatively : It is easier to solve the problem in a coordinate system related to the first motorboat, i.e., the inertial reference frame is fixed with the first boat instead of earth. Then the second boat moves in relation to this reference frame with a relative velocity $\mathbf{v} = \mathbf{v_b} - \mathbf{v_a}$.

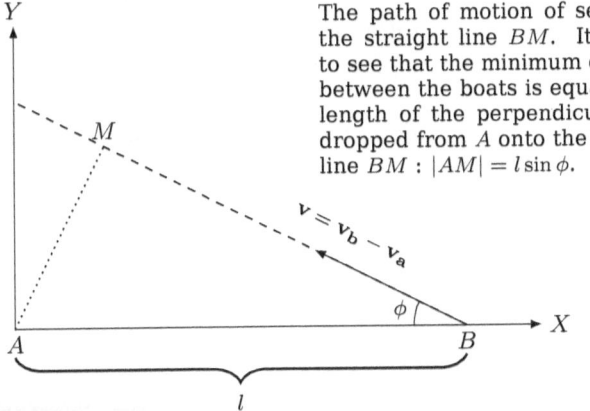

The path of motion of second is the straight line BM. It is easy to see that the minimum distance between the boats is equal to the length of the perpendicular AM dropped from A onto the straight line BM : $|AM| = l \sin\phi$.

Projecting $\mathbf{v} = \mathbf{v_b} - \mathbf{v_a}$ on the Y-axis

$$v \sin\phi = v_b \sin\beta - v_a \sin\alpha.$$

By the law of cosines,

$$v = \sqrt{v_a^2 + v_b^2 + 2v_a v_b \cos(\alpha+\beta)}.$$

$$\therefore \sin\phi = \frac{v_b \sin\beta - v_a \sin\alpha}{\sqrt{v_a^2 + v_b^2 + 2v_a v_b \cos(\alpha+\beta)}}.$$

$$\therefore r_{min} = |AM| = \frac{l(v_b \sin\beta - v_a \sin\alpha)}{\sqrt{v_a^2 + v_b^2 + 2v_a v_b \cos(\alpha+\beta)}}.$$ ∎

§ Problem 2.2.3.
A motorboat travels from port A to B at a speed v_α with respect to the water. At the same time another motorboat leaves B for A at a speed v_β. While the first boat moves between the ports, the second boat covers this distance n times and reaches B at the same time as the first boat. Determine the velocity of the current. ◊

§§ Solution. Let us assume that the river flows from A to B with a velocity of v_0 and l is the distance between the ports.

Since the duration of motion of the boats is the same:
$$\therefore \frac{l}{v_\alpha + v_0} = \frac{n}{2}\left(\frac{l}{v_\beta + v_0} + \frac{l}{v_\beta - v_0}\right) = \frac{nlv_\beta}{v_\beta^2 - v_0^2},$$
$$\therefore v_0^2 + (nv_\beta)v_0 + (nv_\alpha v_\beta - v_\beta^2) = 0$$
$$\therefore v_0 = -\frac{n}{2}v_\beta \pm \sqrt{\left(1 + \frac{n^2}{4}\right) - nv_\alpha v_\beta}.$$ ∎

§ Problem 2.2.4.
Two boats, A and B, move away from a buoy anchored at the middle of a river along the mutually perpendicular straight lines: the boat A along the river, and the boat B across the river. Having moved off an equal distance from the buoy the boats returned. Find the ratio of times of motion of boats $\frac{\tau_A}{\tau_B}$ if the velocity of each boat with respect to water is η times greater than the stream velocity. ◊

§§ Solution. The stream velocity is v_0 and the velocity of each boat with respect to water is v_b.

The buoy is anchored at O.

Distance $OA = OB = l$.

Then, the time of motion of boat A =
$$\tau_A = \tau_{OA} + \tau_{AO}$$
$$= \frac{l}{v_b + v_0} + \frac{l}{v_b - v_0}$$
$$= \frac{2lv_b}{v_b^2 - v_0^2}.$$

Since the path of the motion of boat B is always the straight line OB, hence the components of the velocities of the current and of the boat in the direction perpendicular to OB should be equal, i.e., if it heads at a certain angle α to the straight line OB against the current, then its velocity along the river is zero, i.e., $v_s \sin\alpha = v_0$.

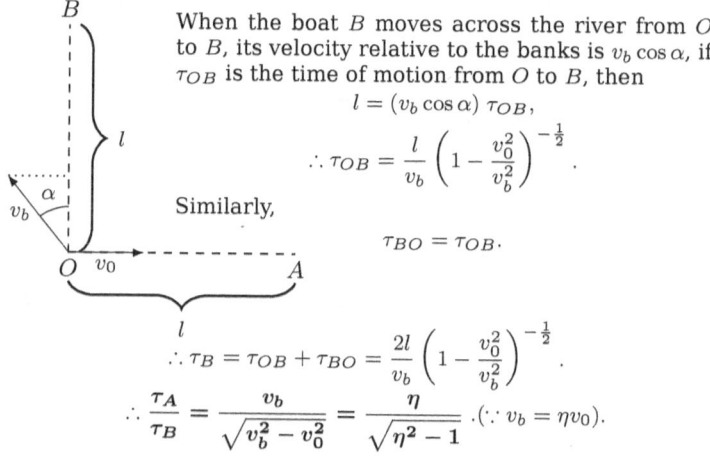

When the boat B moves across the river from O to B, its velocity relative to the banks is $v_b \cos\alpha$, if τ_{OB} is the time of motion from O to B, then
$$l = (v_b \cos\alpha)\,\tau_{OB},$$
$$\therefore \tau_{OB} = \frac{l}{v_b}\left(1 - \frac{v_0^2}{v_b^2}\right)^{-\frac{1}{2}}.$$

Similarly,
$$\tau_{BO} = \tau_{OB}.$$

$$\therefore \tau_B = \tau_{OB} + \tau_{BO} = \frac{2l}{v_b}\left(1 - \frac{v_0^2}{v_b^2}\right)^{-\frac{1}{2}}.$$

$$\therefore \frac{\tau_A}{\tau_B} = \frac{v_b}{\sqrt{v_b^2 - v_0^2}} = \frac{\eta}{\sqrt{\eta^2 - 1}}. \quad (\because v_b = \eta v_0).$$ ∎

2.2. Rendezvous with Anchored Ports

§ Problem 2.2.5. *A wind is blowing with a constant velocity v_w in the direction denoted by the arrow in the figure.*

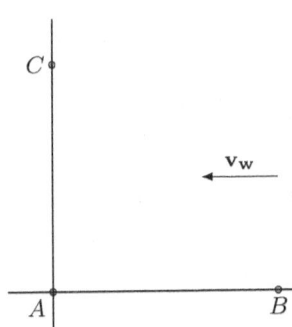

Two airplanes start out from a point A and fly with a constant speed v_p. One flies against the wind to a point B and then returns to point A, while the other flies in the direction perpendicular to the wind to a point C and then returns to point A.

The distances AB and AC are the same. Which plane will return to point A first and what will be the ratio of the flight times of the two planes? ◊

§§ Solution. $AB = AC = l$ (say).

Total time of flight of the first plane =
$$\tau_\alpha = \tau_{AB} + \tau_{BA} = \frac{l}{v_p - v_w} + \frac{l}{v_p + v_w} = \frac{2lv_p}{v_p^2 - v_w^2}.$$

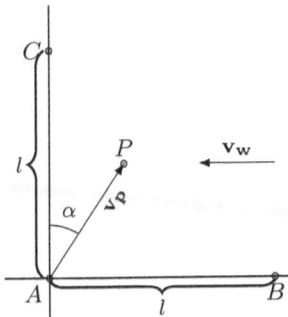

Since the path of the motion of the second plane is always the straight line AC, hence the components of the velocities of the wind and of the plane in the direction perpendicular to AC should be equal, i.e., if it heads at a certain angle α to the straight line AC against the wind, then its velocity along AB is zero, i.e., $v_p \sin \alpha = v_w$.

When the second plane moves across the wind from A to C, its velocity along AC is $v_p \cos \alpha$.

If τ_{AC} is the time of motion from A to C, then
$$l = (v_p \cos \alpha) \tau_{AC}.$$
$$\therefore \tau_{AC} = \frac{l}{v_p \sqrt{1 - \frac{v_w^2}{v_p^2}}} = \frac{l}{\sqrt{v_p^2 - v_w^2}}.$$

Similarly,
$$\tau_{CA} = \tau_{AC}.$$

Total time of flight of the second plane =
$$\tau_\beta = \tau_{AC} + \tau_{CA} = \frac{2l}{\sqrt{v_p^2 - v_w^2}}.$$
$$\therefore \frac{\tau_\alpha}{\tau_\beta} = \frac{v_p}{\sqrt{v_p^2 - v_w^2}} > 1,$$
$$\therefore \tau_\alpha > \tau_\beta.$$

∴ **The second plane will return to point A first.** ∎

2.3 Crossing The River

§ Problem 2.3.1.

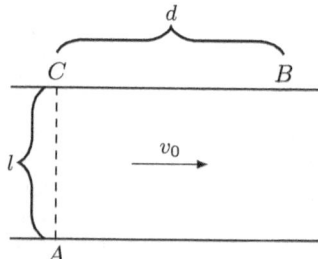

A motorboat must get from point A to point B on the opposite bank of the river. The distance $BC = d$. The width of the river $AC = l$. Assuming the current velocity v_0, determine the minimum speed of the boat relative to the water. ◊

§§ Solution. The speed of the boat **v** with respect to the bank is directed along AB.

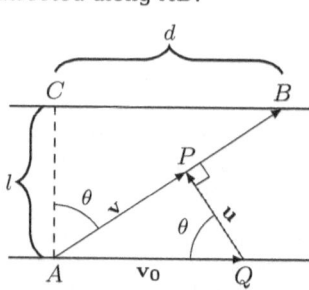

Hence, $\mathbf{v} = \mathbf{v_0} + \mathbf{u}$.

We know the direction of vector **v** and the magnitude and direction of vector $\mathbf{v_0}$.

As can be seen in the drawing that vector **u** will be minimum when $\mathbf{u} \perp \mathbf{v}$.

$\therefore u_{min} = v_0 \cos\theta,$

where $\cos\theta = \dfrac{l}{\sqrt{d^2 + l^2}}.$ ∎

§ Problem 2.3.2.

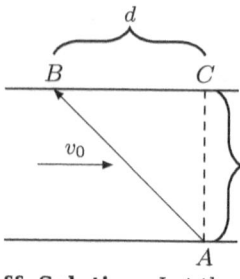

A motorboat must get from point A on one bank of a river to point B on the other bank moving along straight line AB.

The width of the river $AC = l$, the distance $BC = d$, the maximum speed of the boat relative to the water v_b and the river current velocity v_0, determine the minimum time to cover the distance AB? ◊

§§ Solution. Let the speed of the boat v_b be directed at an angle α to the bank.

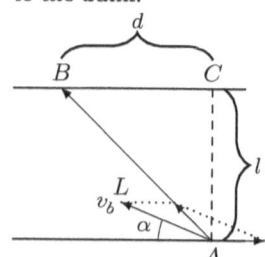

$$BC = d = \tau(v_b \cos\alpha - v_0),$$
$$AC = l = \tau\, v_b \sin\alpha,$$

where τ is the time the boat is in motion.

$\therefore 1 = \sin^2\alpha + \cos^2\alpha = \dfrac{l^2}{\tau^2 v_b^2} + \dfrac{1}{v_b^2}\left(v_0 + \dfrac{d}{\tau}\right)^2$

$\therefore \left(v_b^2 - v_0^2\right)\tau^2 - (2v_0 d)\tau - \left(l^2 + d^2\right) = 0.$

2.3. Crossing The River

$$\therefore \tau_{min} = \frac{v_0 d - \sqrt{v_b^2 d^2 + \left(v_b^2 - v_0^2\right)}}{\left(v_b^2 - v_0^2\right)}.$$ ∎

§ Problem 2.3.3. *A motorboat travels across a river from point A to point B on the opposite bank along a straight line AB forming an angle α with the bank.*

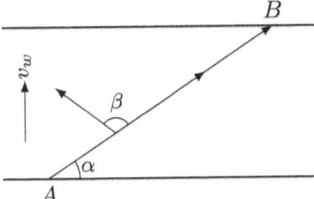

The wind blows with a velocity of v_w at right angles to the bank. The flag on the mast of the boat forms an angle β with the direction of its motion. Determine the speed of the boat with respect to the bank. ◊

§§ Solution. Let $\mathbf{v_{wb}}$ be the velocity of the wind relative to the motorboat. Hence the flag on the mast will be directed along $\mathbf{v_{wb}}$.

If v_b is the speed of the motorboat with respect to the bank, then $\mathbf{v_w} = \mathbf{v_b} + \mathbf{v_{wb}}$.

In $\triangle OFN$:
$$\angle ONF = \beta - \left(\frac{\pi}{2} - \alpha\right)$$
$$= \alpha + \beta - \frac{\pi}{2},$$
$$\angle FON = \angle ONA = \pi - \beta.$$

$$\therefore \frac{OF}{\sin \angle ONF} = \frac{NF}{\sin \angle FON} \text{ (Sine Theorem)}$$

$$\therefore \frac{v_b}{\sin\left(\alpha + \beta - \frac{\pi}{2}\right)} = \frac{v_w}{\sin(\pi - \beta)},$$

$$\therefore v_b = -v_w \frac{\cos(\alpha + \beta)}{\sin \beta}.$$ ∎

§ Problem 2.3.4.

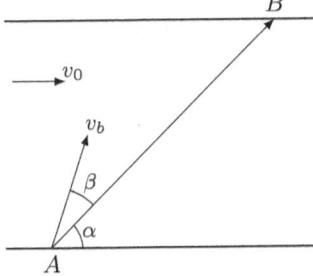

A motorboat moves between two points A and B on the opposite banks of a river, always following the straight line AB.

The distance between points A and $B = s$.

The velocity of the river current v_0 is constant over the entire width of the river. The line AB makes an angle α with the direction of the current.

Determine the velocity v_b and the angle β to the line AB of the boat's motion to cover the distance AB and back in a time τ. The angle β remains the same during the passage from A to B and from B to A. ◊

2.3. Crossing The River

§§ Solution. It is easier to split the velocities of the river current and of the motorboat into components along the line AB and perpendicular to it.

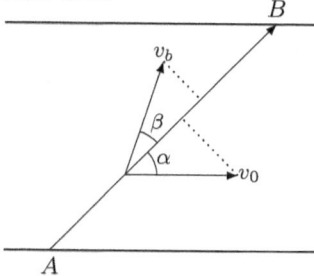

Since the path of the boat's motion is always the straight line AB, hence the components of the velocities of the current and of the boat in the direction perpendicular to AB should be equal:

$$\therefore v_b \sin\beta = v_0 \sin\alpha.$$

When the boat moves from A to B, its velocity relative to the banks is $v_b\cos\beta + v_0\cos\alpha$, if τ_{AB} is the time of motion, then

$$s = (v_b\cos\beta + v_0\cos\alpha)\,\tau_{AB}$$

When the boat moves from B to A, its velocity relative to the banks is $v_b\cos\beta - v_0\cos\alpha$, if τ_{BA} is the time of motion, then

$$s = (v_b\cos\beta - v_0\cos\alpha)\,\tau_{BA}.$$

$$\therefore \tau = \tau_{AB} + \tau_{BA} = \frac{s}{v_b\cos\beta + v_0\cos\alpha} + \frac{s}{v_b\cos\beta - v_0\cos\alpha}$$

$$\therefore \left(v_0^2\tau^2\sin^2\alpha\right)\cot^2\beta - (2sv_0\tau\sin\alpha)\cot\beta - v_0^2\tau^2\cos^2\alpha = 0$$

$$\therefore \cot\beta = \frac{s \pm \sqrt{s^2 + v_0^2\tau^2\cos^2\alpha}}{v_0\tau\sin\alpha}$$

$$\therefore \beta = \cot^{-1}\frac{s \pm \sqrt{s^2 + v_0^2\tau^2\cos^2\alpha}}{v_0\tau\sin\alpha}.$$

$$\therefore v_b = v_0\frac{\sin\alpha}{\sin\beta} = v_0\sin\alpha\sqrt{1+\cot^2\beta}$$

$$\therefore v_b = \frac{\sqrt{v_0^2\tau^2 + 2s\left(s \pm \sqrt{s^2 + v_0^2\tau^2\cos^2\alpha}\right)}}{\tau}.$$ ∎

§ Problem 2.3.5. *A man in a motorboat crosses a river from point A. If he rows perpendicular to the banks then τ_α time after he starts, he will reach point C lying at a distance d downstream from point B.*

2.3. Crossing The River

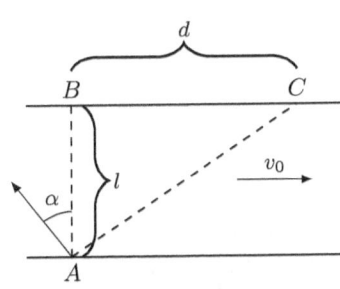

If the man heads at a certain angle α to the straight line AB (AB is perpendicular to the banks) against the current he will reach point B after τ_β time.

Assume the velocity of the boat relative to the water to be constant and of the same magnitude in both cases.

Determine
(a) the width l of the river,
(b) the velocity of the boat v_b relative to the water,
(c) the speed of the current v_0, and
(d) the angle α. ◊

§§ Solution. In both cases the motion of the boat is composed of its motion relative to the water and its motion together with the water relative to the banks.

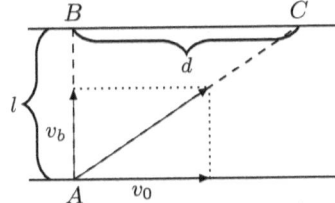

If the boatman rows perpendicular to the banks then the boat moves *along* the river with the speed of the current v_0.
$$\therefore d = v_0 \tau_\alpha.$$

The boat moves *across* the river with a velocity v_b.
$$\therefore l = v_b \tau_\alpha.$$
$$\therefore \frac{v_0}{v_b} = \frac{d}{l}.$$

If the man heads at a certain angle α to the straight line AB against the current, then the velocity of the boat *along* the river is zero,

i.e., $v_b \sin \alpha = v_0$. And $l = (v_b \cos \alpha) \tau_\beta$.
$$\therefore \sin \alpha = \frac{d}{l}, \quad \cos \alpha = \frac{\tau_\alpha}{\tau_\beta}.$$

(a) $\because \sin^2 \alpha + \cos^2 \alpha = 1$,
$$\therefore \frac{d^2}{l^2} + \frac{\tau_\alpha^2}{\tau_\beta^2} = 1,$$
$$\therefore l = \frac{d \tau_\beta}{\sqrt{\tau_\beta^2 - \tau_\alpha^2}}.$$

(b) $v_b = \dfrac{l}{\tau_\alpha} = \dfrac{d \tau_\beta}{\tau_\alpha \sqrt{\tau_\beta^2 - \tau_\alpha^2}}.$

(c) $v_0 = \dfrac{d}{\tau_\alpha}$.

(d) $\alpha = \sin^{-1}\dfrac{d}{l} = \sin^{-1}\dfrac{\sqrt{\tau_\beta^2 - \tau_\alpha^2}}{\tau_\beta}$. ∎

§ **Problem 2.3.6.** *From a point A on a bank of a river with still waters a person must get to a point B situated at a distance b from a point E on the opposite bank. Width of the river $AF = l$, the distance $EF = d$. The person uses a motorboat to travel across the channel and then walks along the bank to point B.*

The velocity of the boat is v_b and the velocity of the walking person is v_p. Determine the minimum time for the person to get from A to B. The angles α and β are as shown.

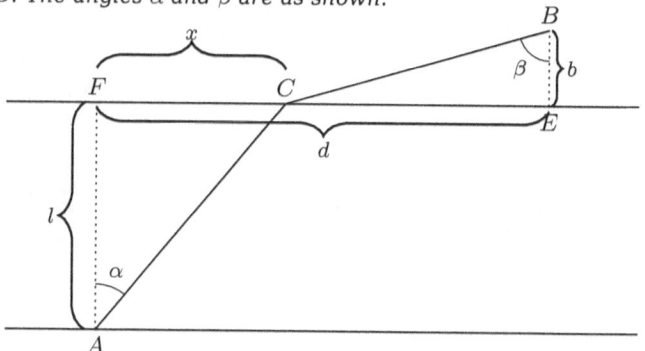

◊

§§ **Solution.** Time of travel from A to $C =$
$$\tau_{AC} = \dfrac{AC}{v_b} = \dfrac{\sqrt{l^2 + x^2}}{v_b}.$$
Time of travel from C to $B =$
$$\tau_{CB} = \dfrac{CB}{v_p} = \dfrac{\sqrt{b^2 + (d-x)^2}}{v_p}.$$
Hence, the total time of travel from A to $B =$
$$t = \tau_{AC} + \tau_{CB} = \dfrac{\sqrt{l^2 + x^2}}{v_b} + \dfrac{\sqrt{b^2 + (d-x)^2}}{v_p}.$$
Extrema of x lies in the solution of
$$\dfrac{dt}{dx} = 0,$$
$$\therefore \dfrac{x}{v_b\sqrt{l^2+x^2}} - \dfrac{(d-x)}{v_p\sqrt{b^2+(d-x)^2}} = 0.$$
$$\therefore \dfrac{\sin\alpha}{v_b} = \dfrac{\sin\beta}{v_p},$$
$$\therefore \dfrac{\sin\alpha}{\sin\beta} = \dfrac{v_b}{v_p}.$$
$$\therefore t_{min} = \dfrac{l^2 + x^2\tan^2\alpha}{v_b} + \dfrac{b^2 + b^2\tan^2\beta}{v_p}$$

2.3. Crossing The River

$$= \frac{l}{v_b \cos \alpha} + \frac{b}{v_p \cos \beta}$$
$$= \frac{l}{v_b \sqrt{1 - \sin^2 \alpha}} + \frac{b}{v_p \sqrt{1 - \sin^2 \beta}},$$

where $\dfrac{\sin \alpha}{\sin \beta} = \dfrac{v_b}{v_p}$. ∎

§ Problem 2.3.7. *A man is on the shore of a lake at point A, and has to get to point B on the lake in the shortest possible time.*

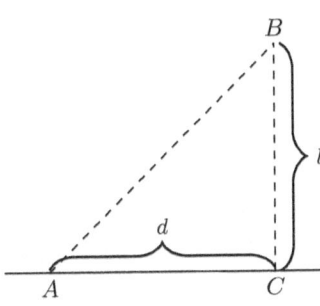

The distance from point B to the shore $BC = l$ and the distance $AC = d$. The man can swim in the water with a speed of v_s and run along the shore with a speed of v_r, greater than v_s.

Determine the path he should use: swim from point A straight to B or run a certain distance along the shore and then swim to point B? ◇

§§ Solution. Assume that the man follows the path ADB where $DC = x$, $\angle DBC = \alpha$.

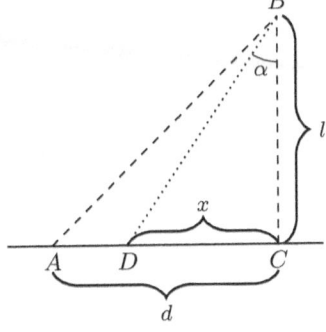

The time of motion
$$t = \frac{d - x}{v_r} + \frac{\sqrt{l^2 + x^2}}{v_s}.$$

$$\therefore \frac{dt}{dx} = 0 = -\frac{1}{v_r} + \frac{x}{v_s \sqrt{l^2 + x^2}},$$
$$\therefore \frac{x}{\sqrt{l^2 + x^2}} = \sin \alpha = \frac{v_s}{v_r}.$$
$$\therefore x = \frac{l v_s}{\sqrt{v_r^2 - v_s^2}}.$$

This corresponds to t_{min} $\because \dfrac{d^2 t}{dx^2} > 0.$

$$\therefore \frac{d^2 t}{dx^2} = \frac{1 + 2x^2}{v_s \sqrt{l^2 + x^2}} = \frac{v_r^2 - v_s^2 + 2 v_s^2 l^2}{l v_r v_s \sqrt{v_r^2 - v_s^2}} > 0 \ (\because v_r > v_s)$$

2.3. Crossing The River

Hence if $d \leq \dfrac{lv_s}{\sqrt{v_r^2 - v_s^2}}$, the man should immediately swim to point B along AB. Otherwise, the man should run along the shore over the distance $AD = d - \dfrac{lv_s}{\sqrt{v_r^2 - v_s^2}}$ and then swim to B.

Alternatively, Since the man's speed in water is lower than that along the shore, the route AB will not necessarily take the shortest time.

The time of motion
$$t = \frac{d-x}{v_r} + \frac{\sqrt{l^2+x^2}}{v_s}$$
$$= \frac{\left(v_r\sqrt{l^2+x^2} - v_s x\right) + v_s d}{v_r v_s}.$$

This time will be minimum if
$$y = v_r\sqrt{l^2+x^2} - v_s x$$
has the smallest value.

Obviously, the value of x that corresponds to the minimum time t does not depend on the distance d.

To find the value of x corresponding to the minimum value of y, expressing x through y the quadratic equation follows
$$x^2 - \frac{2yv_s}{v_r^2 - v_s^2}x + \frac{v_r^2 l^2 - y^2}{v_r^2 - v_s^2} = 0.$$
$$\therefore x = \frac{v_s y \pm v_r\sqrt{y^2 + l^2 v_s^2 - l^2 v_r^2}}{v_r^2 - v_s^2}$$

For x to be real,
$$y^2 + l^2 v_s^2 \geq l^2 v_r^2,$$
$$\therefore y_{min} = l\sqrt{v_r^2 - v_s^2}$$
which corresponds to $x = \dfrac{lv_s}{\sqrt{v_r^2 - v_s^2}}$.

Hence if $d \leq \dfrac{lv_s}{\sqrt{v_r^2 - v_s^2}}$, the man should immediately swim to point B along AB. Otherwise, the man should run along the shore over the distance $AD = d - \dfrac{lv_s}{\sqrt{v_r^2 - v_s^2}}$ and then swim to B. ∎

§ **Problem 2.3.8.** *Two swimmers leave point A on one bank of the river to reach point B lying right across on the other bank. One of them crosses the river along the straight line AB while the other swims at right angles to the stream and then walks the distance that he has been carried away by the stream to get to point B. What was the velocity u of his walking if both swimmers reached the destination simultaneously ? The stream velocity is v_0 and the velocity of each swimmer with respect to water is v_s.* ◊

§§ **Solution.** In both cases the motion of the swimmer is composed of its motion relative to the water and its motion together with the

2.3. Crossing The River

water relative to the banks. It is easier to split the velocities of the river current and of the swimmers into components along the line AB and perpendicular to it. Width of the river is $AB = l$.

Since the path of the first swimmer's motion is always the straight line AB, hence the components of the velocities of the current and of the swimmer in the direction perpendicular to AB should be equal,

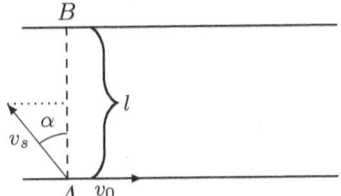

i.e., if he heads at a certain angle α to the straight line AB against the current, then his velocity along the river is zero, i.e.,
$$v_s \sin \alpha = v_0.$$

When the swimmer crosses the river from A to B, its velocity relative to the banks is $v_s \cos \alpha$, if τ_{AB} is the time of motion, then
$$l = (v_s \cos \alpha)\, \tau_{AB},$$
$$\therefore \tau_{AB} = \frac{l}{v_s} \left(1 - \frac{v_0^2}{v_s^2}\right)^{-\frac{1}{2}}.$$

For the second swimmer, his motion is perpendicular to the banks and he gets drifted along the river to C with the speed of the current v_0,
$$\therefore d = v_0 \tau_{AC}.$$

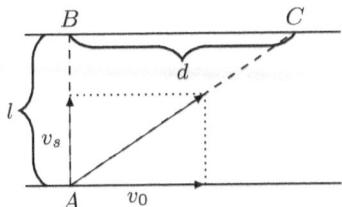

He swims across the river with a velocity v_s,
$$\therefore l = v_s \tau_{AC},$$
$$\therefore \tau_{AC} = \frac{l}{v_s}.$$
And, the time of walking
$$\tau_{CB} = \frac{d}{u} = \frac{lv_0}{uv_s}.$$

$$\because \tau_{AB} = \tau_{AC} + \tau_{CB},$$
$$\therefore \frac{l}{v_s}\left(1 - \frac{v_0^2}{v_s^2}\right)^{-\frac{1}{2}} = \frac{l}{v_s} + \frac{lv_0}{uv_s},$$
$$\therefore u = \frac{v_0}{\left(1 - \dfrac{v_0^2}{v_s^2}\right)^{-\frac{1}{2}} - 1}. \qquad \blacksquare$$

§ Problem 2.3.9. *A boat moves relative to water with a velocity which is n times less than the river flow velocity. At what angle to the stream direction must the boat move to minimize drifting? Determine the minimum drift distance, width of the river is l.* ◊

2.3. Crossing The River

§§ Solution. The stream velocity is v_0 and the velocity of the boat with respect to water is v_b. Width of the river $AB = l$.
Drift distance $BC = x$.

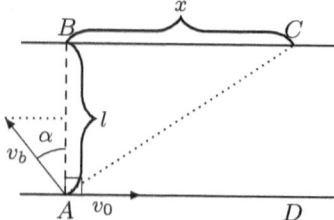

If the boat heads at a certain angle α to the straight line AB against the current, then his velocity along the river is $v_0 - v_b \sin \alpha$ and its velocity across the river (relative to the banks) is $v_b \cos \alpha$.

$$\therefore l = (v_b \cos \alpha)\tau,$$

And
$$x = (v_0 - v_b \sin\alpha)\tau = l\frac{v_0}{v_b}\left(\sec\alpha - \frac{v_b}{v_0}\tan\alpha\right),$$

where τ is the time of motion.

Extrema of x is found by $\dfrac{dx}{d\alpha} = 0$,

$$\therefore \left(\sec\alpha \tan\alpha - \frac{v_b}{v_0}\sec^2\alpha\right) = 0,$$

$$\therefore \sin\alpha = \frac{v_b}{v_0} = \frac{1}{n}.$$

$$\frac{d^2 x}{d^2 \alpha} = \frac{n}{\sqrt{n^2-1}} > 0.$$

$\therefore \alpha = \sin^{-1}\dfrac{1}{n}$ corresponds to x_{min}.

Hence the required angle to the stream direction AD =
$$\frac{\pi}{2} + \alpha = \sin^{-1}\frac{1}{n} + \frac{\pi}{2}.$$

$$x_{min} = l\frac{v_0}{v_b}\left(\sec\alpha - \frac{v_b}{v_0}\tan\alpha\right)$$
$$= nl\left(\frac{n}{\sqrt{n^2-1}} - \frac{1}{n\sqrt{n^2-1}}\right)$$
$$= l\sqrt{n^2-1}.$$

Alternatively:

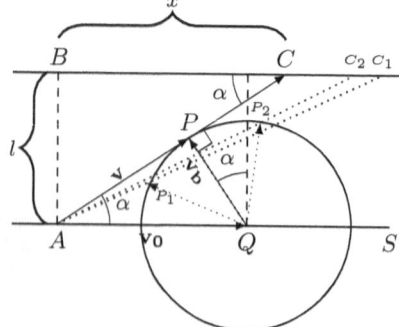

The speed of the boat **v** with respect to the bank is directed along AC. Hence, **v** = **v₀** + **v_b**. As can be seen in the drawing that the drift distance BC is minimum with AP being tangent to the circle with radius $QP = QP_1 = QP_2$. $\therefore$ **v_b** $\perp$ **v**.

In $\triangle APQ$:
$$\sin\alpha = \frac{PQ}{AQ} = \frac{v_b}{v_0} = \frac{1}{n}.$$
Hence the required angle to the stream direction $AS =$
$$\angle PQS = \frac{\pi}{2} + \alpha = \sin^{-1}\frac{1}{n} + \frac{\pi}{2}.$$
In $\triangle ABC$:
$$\tan\alpha = \frac{AB}{BC} = \frac{l}{x},$$
$$\therefore x = l\sqrt{n^2 - 1}.$$
■

§ **Problem 2.3.10.** *From point A located on a highway one has to get by car as soon as possible to point B located in the field at a distance l from the highway.*

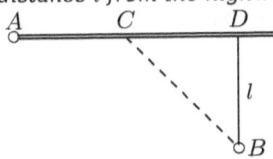

It is known that the car moves in the field η times slower than on the highway. At what distance from point D one must turn off the highway? ◇

§§ **Solution**. Assume that the path followed is ACB where $CD = x$, $AD = d$.

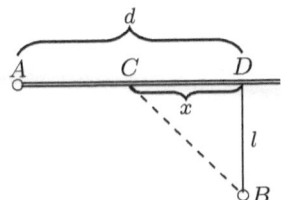

The time of motion
$$t = \frac{AC}{v_h} + \frac{CB}{v_f}$$
$$= \frac{d - x}{v_h} + \frac{\sqrt{l^2 + x^2}}{v_f}.$$

Where v_h is the speed of car on highway and v_f is the speed of car on field, and $v_f = \frac{v_h}{\eta}$.

Extrema of t is found by $\frac{dt}{dx} = 0$
$$\therefore \frac{1}{v_h}\left(-1 + \eta\frac{x}{\sqrt{l^2 + x^2}}\right) = 0,$$
$$\therefore x = \frac{l}{\sqrt{\eta^2 - 1}}.$$

This corresponds to t_{min} ∵ $\frac{d^2t}{dx^2} > 0$

$$\therefore \frac{d^2t}{dx^2} = \frac{1 + 2x^2}{v_f\sqrt{l^2 + x^2}} = \frac{2l^2 + \eta^2 - 1}{\eta l\sqrt{\eta^2 - 1}} > 0 \;(\because \eta > 1)$$

2.3. Crossing The River

Hence if $d \le \dfrac{l}{\sqrt{\eta^2 - 1}}$, one should immediately turn off the highway at A itself and move in the field to point B along AB. Otherwise, one should move on the highway over the distance $AC = d - \dfrac{l}{\sqrt{\eta^2 - 1}}$ and then move in the field to B along CB.

Alternatively, Since the car's speed in field is lower than that on the highway, the route AB will not necessarily take the shortest time.

The time of motion
$$t = \frac{d-x}{v_h} + \frac{\sqrt{l^2+x^2}}{v_f} = \frac{\left(v_h\sqrt{l^2+x^2} - v_f x\right) + v_f d}{v_h v_f}.$$

This time will be minimum if $y = v_h\sqrt{l^2+x^2} - v_f x$ has the smallest value. Obviously, the value of x that corresponds to the minimum time t does not depend on the distance d. To find the value of x corresponding to the minimum value of y, expressing x through y the quadratic equation follows

$$x^2 - \frac{2yv_f}{v_h^2 - v_f^2}x + \frac{v_h^2 l^2 - y^2}{v_h^2 - v_f^2} = 0.$$

$$\therefore x = \frac{v_f y \pm v_h\sqrt{y^2 + l^2 v_f^2 - l^2 v_h^2}}{v_h^2 - v_f^2}$$

For x to be real, $y^2 + l^2 v_f^2 \ge l^2 v_h^2$.

$\therefore y_{min} = l\sqrt{v_h^2 - v_f^2}$ which corresponds to

$$x = \frac{lv_f}{\sqrt{v_h^2 - v_f^2}} = \frac{l}{\sqrt{\eta^2 - 1}}. \qquad \blacksquare$$

§ **Problem 2.3.11.** *A boat is moving across a river whose waters flow with a velocity* $\mathbf{v_0}$.

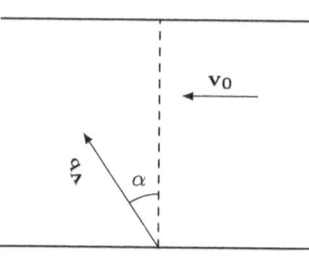

The velocity of the boat with respect to the current, $\mathbf{v_b}$, *is directed at an angle* α *to the line perpendicular to the current.*

Determine

(a) *the angle* θ *at which the boat moves with respect to this line.*

(b) *the velocity* $\mathbf{v}$ *of the boat with respect to the river banks.*

(c) *the angle at which the boat moves directly across the current with given* $\mathbf{v_0}$ *and* $\mathbf{v}$? ◊

§§ **Solution.** (a)
$$\tan\theta = \frac{v_0 + v_b \sin\alpha}{v_b \cos\alpha}$$

$$\therefore \theta = \tan^{-1}\left(\tan\alpha + \frac{v_0}{v_b \cos\alpha}\right).$$

2.3. Crossing The River

(b)
$$v^2 = (v_b \cos\alpha)^2 + (v_0 + v_b \sin\alpha)^2$$
$$\therefore v = v_b\sqrt{1 + 2\frac{v_0}{v_b}\sin\alpha + \left(\frac{v_0}{v_b}\right)^2}.$$

(c) The boat will move directly across the current if $\theta = 0$
$$\therefore \sin\alpha = -\frac{v_0}{v_b},$$
$$\therefore \alpha = \sin^{-1} -\frac{v_0}{v_b}.$$

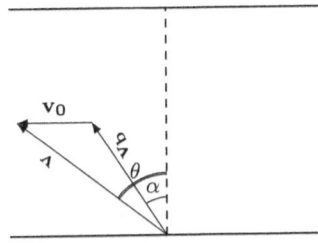

It is easy to see that the boat can travel at right angles to the current only if $v_b > v_0$.

■

§ Problem 2.3.12. *A swimmer crosses a river that has a width h. At what angle and to the direction of flowing he should swim to cross the opposite shore in the shortest time? Determine the path of the swimmer if the speed of the river is equal to v_0, and the speed of the swimmer relative to the water is equal to v_s?* ◊

§§ Solution. The swimmer starts at O and swims at an angle α to OX. The river flows along OX with the speed of v_0.

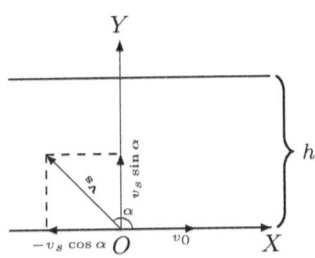

Then the law of motion of the swimmer is given by
$$x = (v_0 + v_s \cos\alpha)t,$$
$$y = (v_s \sin\alpha)t.$$

The swimmer gets to the other shore when $y = h$.

The time of motion $\tau = \dfrac{h}{v_s \sin\alpha}$.

It will be minimal when $\sin\alpha$ is maximal, i.e.
$$\therefore \alpha = \frac{\pi}{2}.$$
$$\therefore \tau_{min} = \frac{h}{v_s}.$$

For $\alpha = \dfrac{\pi}{2}$, we have $x = v_0 t$.

Therefore, when the swimmer will be on the other bank,
$$PC = d = x_{min} = v_0 \tau_{min}$$
$$= \frac{hv_0}{v_s}.$$

2.3. Crossing The River

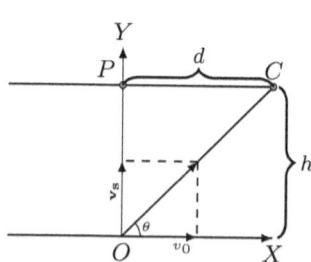

Hence length of the traversed path $OC =$
$$s = \sqrt{d^2 + h^2} = \frac{h}{v_s}\sqrt{v_0^2 + v_s^2}.$$
And,
$$\tan\theta = \frac{h}{d} = \frac{v_s}{v_0},$$
$$\therefore \theta = \tan^{-1}\frac{v_s}{v_0}.$$

∎

§ Problem 2.3.13.
A boatman crosses a river of width h from point A to point B, all the time directing the boat at an angle α to the shore.

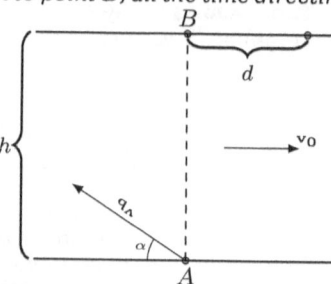

Find the speed of the boat v_b relative to the water, if the speed of the river is equal to v_0, and the boat is drifted below point B by a distance d.

◊

§§ Solution.
Laws of motion is given by :

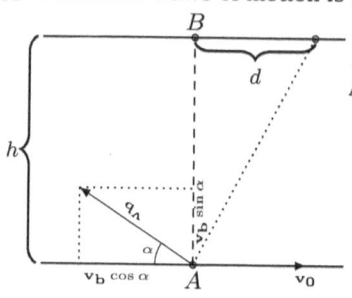

$$d = (v_0 - v_b \cos\alpha)\tau,$$
And
$$h = (v_b \sin\alpha)\tau.$$
$$\therefore v_b = \frac{hv_0}{d\sin\alpha + h\cos\alpha}.$$

∎

www.ingramcontent.com/pod-product-compliance
Lightning Source LLC
Chambersburg PA
CBHW031548210526
45464CB00003B/1200